Y023689

D1611106

Charles Darwin's Debt
to the Romantics

Charles Darwin's Debt to the Romantics

How Alexander von Humboldt, Goethe and
Wordsworth Helped Shape Darwin's View of Nature

Charles Morris Lansley

PETER LANG

Oxford • Bern • Berlin • Bruxelles • New York • Wien

Bibliographic information published by Die Deutsche Nationalbibliothek.
Die Deutsche Nationalbibliothek lists this publication in the Deutsche
Nationalbibliografie; detailed bibliographic data is available on the
Internet at http://dnb.d-nb.de.

A catalogue record for this book is available from the British Library.

Library of Congress Control Number: 2018933752

Cover image © Julia Dobson Design and Illustration.

ISBN 978-1-78707-138-4 (print) • ISBN 978-1-78707-139-1 (ePDF)
ISBN 978-1-78707-140-7 (ePub) • ISBN 978-1-78707-141-4 (mobi)

© Peter Lang AG 2018

Published by Peter Lang Ltd, International Academic Publishers,
52 St Giles, Oxford, OX1 3LU, United Kingdom
oxford@peterlang.com, www.peterlang.com

Charles Morris Lansley has asserted his right under the Copyright, Designs
and Patents Act, 1988, to be identified as Author of this Work.

All rights reserved.
All parts of this publication are protected by copyright.
Any utilisation outside the strict limits of the copyright law, without
the permission of the publisher, is forbidden and liable to prosecution.
This applies in particular to reproductions, translations, microfilming,
and storage and processing in electronic retrieval systems.

This publication has been peer reviewed.

Printed in Germany

Contents

List of Figures — vii

Acknowledgements — xi

INTRODUCTION
Charles Darwin's Victorian Debt to the Romantics — 1

CHAPTER 1
Organic and One Reality Nature in Humboldt and Darwin — 19

CHAPTER 2
The Forces of Nature in Humboldt and Darwin — 51

CHAPTER 3
Darwin's Romantic Theory of Nature — 63

CHAPTER 4
Darwin's Romantic Theory of Mind — 81

CHAPTER 5
Darwin's Concepts of Morality and Romantic Materialism — 119

CHAPTER 6
Darwin's Moral and Reflective Nature: Conflicting Values in the Victorian Era — 165

CHAPTER 7
The Transmutation of Darwin's Romanticism — 191

CHAPTER 8
From Erasmus Darwin's Broth of Chaos to his Goddess of Nature   221

CHAPTER 9
The Rime of the Ancient Naturalist   239

Bibliography   255

Index   267

Figures

Figure 1: A. v. Humboldt, *Geography of the Plants near the Equator*, 1803. © Museo Nacional de Colombia/ Oscar Monsalve Pino. Permission to reproduce the photo kindly granted by Museo Nacional de Colombia and Oscar Monsalve Pino. Colección Museo Nacional de Colombia, reg. 1204.
Alexander von Humboldt (1769/1859). Geografia de las plantas cerca del Ecuador.
Tabla fisica de los Andes y paises vecinos, levantada sobre las observaciones y medidas tomadas en los lugares en 1799–1803, 1803. Acuarela (Acuarela y tinta Papel) 38.7 × 50.3 cm.

Figure 2: L. A. Schönberger and P. J. F. Turpin after A. v. Humboldt and A. Bonpland, *Geógraphie des plantes équinoxiales*, 1807. Permission to reproduce image from book kindly granted by Peter H. Raven Library/Missouri Botanical Garden and the Biodiversity Heritage Library <http://www.biodiversitylibrary.org>.
Essai sur la geographie des plantes: accompagne d'un tableau physique des regions equinoxiales, fonde sur des mesures executes, depuis le dixieme degree de latitude boreale jusqu'au dixieme degree de latitude austral, pendant les annees 1799, 1800, 1801, 1802 et 1803/par Al. de Humboldt et A. Bonpland; redigee par Al. de Humboldt.
A Paris, Chez Levrault, Schoell et compagnie, libraires, XIII–1805.
<http://www.biodiversitylibrary.org/bibliography/9309>.

	Item: <http://www.biodiversitylibrary.org/item/37872>, page 156.	97
Figure 3:	Leaf Sequence in Sidalcea Malviflora. From Miller, 2009, Image 55, p. 107. © Gordon L. Miller and The MIT Press, Cambridge, MA, USA. Permission to reproduce the photo kindly granted by Gordon L. Miller and The MIT Press.	107
Figure 4:	'Part of secondary wing-feather of Argus pheasant, shewing two perfect ocelli, *a* and *b*. A,B,C,D, &c., are dark stripes running obliquely down, each to an occelus. [Much of the web on both sides, especially to the left of the shaft, has been cut off.]'. Text and drawing from Figure 57 in Darwin, 2004, p. 489.	157
Figure 5:	'Basal part of the secondary wing-feather [of the Argus pheasant], nearest to the body'. Text and drawing are from Figure 58 in Darwin, 2004, p. 490.	157
Figure 6:	'Portion of one of the secondary wing-feathers [of the Argus pheasant] near to the body, shewing the so-called elliptic ornaments. The right-hand figure is given merely as a diagram for the sake of the letters of reference. A,B,C,D, &c. Rows of spots running down to and forming the elliptic ornaments. *B*. Lowest spot or mark in row B. *c*. The next succeeding spot or mark in the same row. *d*. Apparently a broken prolongation of the spot *c* in the same row B'. Text and drawing are from Figure 59 in Darwin, 2004, p. 491.	158
Figure 7:	'An ocellus in an intermediate condition between the elliptic ornament and the perfect ball-and-socket ocellus'. Text and drawing are from Figure 60 in Darwin, 2004, p. 492.	158

Figure 8:   'Portion near summit of one of the secondary wing-feathers [of the Argus pheasant], bearing perfect ball-and-socket ocelli. *a.* Ornamented upper part. *b.* Uppermost, imperfect ball-and-socket ocellus. (The shading above the white mark on the summit of the ocellus is here a little too dark.). *c.* Perfect occelus'.
Text and drawing are from Figure 61 in Darwin, 2004, p. 494.   159

Figure 9:   *The Raising of Lazarus* (1517–1519) by Sebastiano del Piombo. The image has been reproduced with the kind permission of the National Gallery, London. © The National Gallery, London.   185

Figure 10: *Diana of Ephesus*. This image has been reproduced with kind permission of Whetton & Grosch. Replica sculpture of Diana of Ephesus. © Whetton & Grosch, Museum Models 2016. <http://www.whettonandgrosch.co.uk>.   216

Penguin Group UK (PUK), and the original publishers, John Murray, now part of Hodder and Stoughton Ltd, have no objection to the reproduction of Figures 4–8 in this volume. However, PUK have been unable to trace the original contract or any other rights information regarding the *Descent* title. They are therefore unable to give formal permission to use the images as they cannot warrant that such use would not infringe any third party rights. They therefore request that this disclaimer is entered in the acknowledgements. The original publisher, John Murray, also state that they do not control the copyright and that they no longer have records of the relevant copyright holders. However, they state that the date of publication (1871) suggests that the images may be in the public domain. They request that John Murray is credited as the original publishers. The figures are therefore inserted in this volume in good faith on the assumption that they are out of copyright and that copyright law has not been infringed.

Every effort has been made to trace copyright holders and to obtain their permission for the use of copyright material. The publisher apologizes for any errors or omissions in the above list and would be grateful for notification of any corrections that should be incorporated in future reprints or editions of this book.

Acknowledgements

Firstly I would like to thank my supervisors Dr Gary Farnell (Director of Studies) and Professor Neil Messer for all the help, support and encouragement they have given me for the duration of the research leading up to the PhD thesis as well as their support for the post-doctoral research leading up to the publication of this book. In particular, Gary has helped steer me through the complexities of Romantic and Victorian literature, whilst Neil has helped sharpen my understanding of Creationism and Ethics. Above all, they have inspired me to think creatively without losing my sense of direction. I would also like to thank my external examiners, Professor Tim Fulford and Derek Bunyard, whose input was most helpful. Thanks must also go to Ruth Padel for her permission to use extracts from her biographical poems of Charles Darwin (Padel, 2010), her great-great grandfather, and for the inspiration they gave me in writing the final chapter.

Thanks also to my original supervision team at the University of Gloucestershire, Professor Adam Hart, Professor Shelley Saguaro and Dr Roy Jackson who helped me start out on this venture.

The journey of this research goes right back in time just like Darwin's 'tree of life' and his quest for the origins of life. For me the search for the meaning of life was very much part of the philosophical discussions I had with my brother Peter and our friend Dr Terry Hopton who inspired me to study philosophy at university. The inspiration for developing an enquiring philosophical mind came through my tutors at Oxford, Guy Backus and his wife Professor Irena Backus. This enthusiasm was developed further when studying linguistics for my MA at the University of Southampton under the late Professor Chris Brumfit and his wife Professor Ros Mitchell – this is where I first seriously came across Charles Darwin within the context of the evolution of language. Once I started reading Darwin I could not stop. So thank you Chris and Ros.

As well as seeing Darwin's 'tree of life' in terms of descent, it can also be seen as a 'tree of enchantment' like a Christmas tree covered in baubles, a celebration of life itself and all those privileged to be part of it. So in this sense I would like to thank my late parents, Peter and Ruth Lansley, for enabling me to be a bauble on this tree, and my wife Claire for enabling us to attach our children, Charlotte and William, thereby connecting the past and the present to the future. I would also like to express my indebtedness to my wife Claire for keeping me in the manner whilst carrying out my research.

I would also like to thank my parents-in-law Martyn and Brenda Iffland for accompanying me and Claire on my first visit to Charles and Emma Darwin's home at Down House in 1997 enabling me to purchase my first bundle of Darwin books; and for Martyn finding the 'Darwin Bark' nearby which our friends Ray and Stella Newton framed as a commemoration of our visit. And thanks to Charles Darwin for inspiring me to research his life.

Finally, I would like to dedicate my research to all past, present and future grandparents for making descent possible and for all past, present and future grandchildren for continuing the line of descent from our past progenitors. In particular my maternal grandparents, Rudolf and Ida Kormes, who I never had the privilege of meeting and who died in one of the Nazi concentration camps, and my paternal grandparents, Percy and Ethel Lansley.

To conclude, as a temporary pause at this moment in time, the final dedication goes to our grandson Rupert Morris Maybury, our granddaughter Merryn Isobel Maybury, and to our daughter Charlotte and her husband George Maybury for continuing the line of descent. And to any further offspring that may be born to our children Charlotte and William. In Darwin's own words, 'from so simple a beginning endless forms most beautiful and most wonderful have been, and are being, evolved' (Darwin, [1859] 1985, p. 460).

INTRODUCTION

Charles Darwin's Victorian Debt to the Romantics

The works of Charles Darwin have mainly been explored within the context of the Victorian period and of Darwin as a significant 'Victorian'. There is much less research available, however, that examines Darwin's own intellectual precursors and cultural influences and in particular the Romantic influences on his life.

Today the academic establishment recognizes Darwin's place in the 'history of ideas' as a 'high Victorian'; he is acknowledged by Desmond and Moore (2009, pp. 449–50) as having begun a paradigm shift away from Paley's concept of divine creation of Nature[1] to one of slow 'transformation' or evolution over millions of years. This book will examine how some of the key ideas of the Romantic era may have influenced the development of Darwin's thinking.

Before considering what might have influenced Darwin, it is important to first understand where the term 'Romantic' comes from. The Romantic era is generally thought to cover the period 1770–1870, and the movement's strongholds were mainly in England (Wordsworth, Coleridge and Blake), Scotland (Scott and Burns) and Germany (Friedrich Schlegel, Goethe, Schelling and Schiller), influenced by the writings of the French Rousseau (Melani, 2009, pp. 1–6). Whereas *Naturphilosophie*[2] was concerned with the organic core of nature and its relationship to mind, 'Romanticism' was *Naturphilosophie* with the addition of aesthetic and moral features. Friedrich Schlegel coined the term *romantisch* 'to indicate a specific kind of poetic and morally valued literature' which 'distinguishes a type of science

---

[1] The view that just as a watch indicates its design by a watchmaker, so too does the human eye indicate its design by a Deity.
[2] Natural philosophy, in German.

that retains this aesthetic and moral heritage' (Richards, 2002, pp. 6–11). This distinction is essential to understanding an 'aesthetic' and 'moral' analysis of Darwin's work if Darwin's biology is to be seen as 'Romantic', and, therefore, not mechanistic. Kreis (2009a, pp. 1–17) defines 'Romanticism' as a reaction to the Enlightenment,[3] in which Newtonian natural philosophy was considered to have stifled individuality, freedom and creativity, imposing materialism and robotic sameness. The Romantics did not want to be controlled by the established church but saw the Enlightenment's attack on religious superstition as taking away Man's spirituality and morality (Kreis, 2009b, pp. 1–10). Although Darwin and Rousseau's underlying philosophies were different, they shared common fears. Both were afraid of chaos, revolution and a world without structure. For Rousseau, Man was naturally good (divine) and close to his Creator, but had been corrupted by society. Although Darwin was replacing the divine structure of Nature with a self-contained structure through natural selection,[4] he could not go so far as to say there was no God. At most he was an agnostic.[5] Most scholars would no doubt make a clear distinction between Rousseau's 'noble savage' (or Man born with goodness but corrupted by society), and Darwin's natural Man developing his moral social instincts over time through natural selection.[6] Darwin's theory of 'Community Selection' develops the idea of evolving moral instincts that over time become inbuilt in Man through his moral conscience. So the containment required is self-contained. In that sense this instinct could be regarded as divine (if seen as a manifestation of Mind), although Darwin would not describe this as a cause created by God. Although Rousseau was against rationalism because philosophy

---

[3] The Age of the Enlightenment or the Age of Reason was a movement in the seventeenth and eighteenth centuries which promoted scientific thought to counteract superstition and abuses of power by the church and state.
[4] This is the theory formulated in Darwin's *On the Origin of Species* in 1859 in which changes in an organism allow it to adapt to its environment thereby helping it survive and have more offspring.
[5] A term coined by Thomas Huxley.
[6] 'The first foundation or origin of the moral sense lies in the social instincts, including sympathy; and these instincts no doubt were primarily gained, as in the case of the lower animals, through natural selection' (Darwin, 2004, p. 682).

(the sciences) demystified the world through mathematical truths (thereby killing off God), there is a certain irony in that it took Darwin, a scientist, to put the mystery back into Nature – maybe not with a God from the scriptures but another self-contained spiritual entity that made up the laws of Nature. This suggests that certain aspects of Darwin's theory could be regarded as firmly Romantic.

The terms 'Romantic' or 'Romanticism' used in this book refer to a blend of interrelated concepts coming from the German Romantic movement. One of these ideas is that Nature is one reality in the sense that it is its own creator and is not created by an external force or external laws. This idea reflects a unity of Nature in which all the parts are interrelated and make up the whole. This wholeness of Nature includes the concept of Mind, meaning that mind and matter are part of this oneness, merging together the subjective sensory awareness of the observer with the objective world. This allows the scientific mind to be complemented by an aesthetic judgement in which the emotions and the imagination can interpret Nature and penetrate its deep structures. The world is therefore material in terms of the objective experience, and spiritual or mental in the way the experiences are felt by the observer. This form of Romantic materialism, in which mind and matter are one, is developed by Charles Darwin who argues that the reflective mental constructs of sympathy and conscience have developed over time through natural selection. But for Darwin, his most radical idea is that mind has developed from matter.

Darwin's strain of Romanticism was a different way of seeing and interpreting the world, from Darwin the naturalist in the field noting his empirical experiences to Darwin the 'scientist'[7] reflecting on those experiences, drawing up theories based on his perceptions and the perceptions of other naturalists, reflecting on those theories, and then going back to experiencing Nature again through the filters of those theories. As a student at Cambridge, and through his friendships with academics such as

---

7   Terms like 'scientist' and 'botanist' are twenty-first century terms.

John Stevens Henslow,[8] 'botanist' and mineralogist, and Adam Sedgwick,[9] geologist, he socialized in intellectual circles familiar with Romantic concepts such as 'archetype'.

Richard Owen, a comparative anatomist and associate of Darwin's, uses the term 'archetype'[10] as does Darwin in his *Origin*.[11] In the case of Owen's 'archetype of the vertebrata', this is a basic plan of skeleton in which the head, ribs, pelvis and limbs develop from the vertebrae. For Darwin it was 'an ideal primitive form' depicting 'the homologous construction of the limbs throughout the whole class'. Darwin's use of the term 'archetype' demonstrates that there is a tension within its underlying meaning between the view that archetypes are pre-formed genealogically from a primordial form (or primordial forms) or that they are teleological forms seeking perfection. My interpretation of Darwin's use of the concept is that it can be seen to be a complex mix of the two. This interpretation allows for both material genealogically based primordial forms fixed in space and time to also be regarded as evolving teleological forms seeking perfection; forms

---

8   List of letters between Henslow and Darwin during *Beagle* Voyage at Burkhardt (2009a), p. 633.
9   Letter to Darwin from Adam Sedgwick on 18 September 1831 giving advice on what geology books to read as well as reference to Humboldt's 'Personal Narrative' at Burkhardt (2009a), p. 157.
10 Owen, 1847, and 1849. Cited in Richards (2002), footnote 33, p. 528. Richards summarizes Owen's archetype as follows: 'The archetype of the vertebrata, in Owen's construction, was simply a string of vertebrae. According to his theory, different vertebrate skeletons manifested modifications of this basic plan. So for instance, the bones of the head would be regarded as a development of the several anterior vertebrae, and the ribs, pelvis, and limbs as developments of different processes of more posterior vertebrae' (Richards, 2002, p. 528).
11 In his Glossary to the *Origin*, Darwin defines 'Archetypal' as follows: 'Of or belonging to the Archetype, or ideal primitive form upon which all the beings of a group seem to be organised' (Darwin, 1985, p. 464). Darwin refers to it when discussing Classification: 'If we suppose that the ancient progenitor, the archetype as it may be called, of all mammals, had its limbs constructed on the existing general pattern, for whatever purpose they served, we can at once perceive the plain signification of the homologous construction of the limbs throughout the whole class' (Darwin, 1985, p. 416).

that are constructed from Man's[12] understanding of the world and which therefore may be interpreted as belonging to a collective mind, whether this be in the form of God or Nature. This interpretation becomes clearer when considering Goethe's 'Genetic Method' which he uses to understand the 'leaf' archetype. The method demonstrates the mental ability to move to and fro between all the stages of a plant's development, from the young leaf to the more mature leaf, to the flower and to the seed, as well as the mental ability to move between the individual plant forms and their universal archetypes. My inclusion of the 'Genetic Method', therefore, is to highlight the mind's role in seeing the histories of development within an imaginative and poetic context, whether this be genealogical or teleological.

Notable contributions to the discussion of Romantic influences are to be found in Gillian Beer (1983), George Levine (2008 and 2011) and a post-script chapter in the work of Robert Richards (2002). Richards argues that Darwin's Romanticism led him to attribute a moral conscience to human beings and not one of Malthusian selfishness, and posits the view that Darwin was profoundly influenced by the German Romantics, principally Johann Wolfgang von Goethe and Alexander von Humboldt. Gillian Beer sees Darwin's works as an example of 'Romantic materialism' as science can be said to use instruments to understand Nature without destroying its mystery. What is observed and captured through the senses may be material but the pleasure of the experience is Romantic, whether 'optimistic' or 'pessimistic' (Beer, 1983, pp. 244–5). Darwin uses many metaphors in his texts and these reflect Beer's argument that the pleasures of observation are Romantic. The most common metaphor used is 'struggle'. On one level it depicts plants and animals fighting for survival both as

---

12 Although in this century the term 'Man' is not an appropriate term to refer to men and women, it is used in this book to reflect Darwin's concepts and the concepts of his time, as reflected by his work *The Descent of Man*, for example. This book aims to examine Darwin within the context of his own time rather than to examine his text within the context of the present century. The term 'Man', therefore, refers to humanity, humankind, men and women. However, where 'man' refers to males specifically, 'man' is used, for example in the context where Darwin believed men to be superior to women. The lower case 'm' in 'man' is also used in quotes where 'man' occurs in primary and secondary texts.

individuals and as species. The consequent violent destruction of individuals through this struggle is represented by Darwin's pessimistic metaphorical use of 'wedges' breaking up the face of Nature (Darwin, 1985, p. 119). On another level it can be seen as representing the vibrancy of life, regeneration, adaptation, development, mutuality – in fact everything that makes up Darwin's 'entangled bank'[13] of Nature', success and achievement, not just failure. In the textual analysis of Darwin's works, the use of metaphor and narrative is explored in the book to draw out his theory of natural selection. The question of whether Darwin's *Origin* can be regarded as an example of Romantic literature as well as science is also explored (with reference to George Levine, 2008 and 2011). The use of 'narrative' is particularly important as Darwin's story can be seen as a story of the evolution of life over time. Beer (1983) compares this to the writing of a plot in a novel. Darwin's use of the term 'entangled bank' reflects both the struggles of individuals and species as well as the varied forms of life that reproduce themselves and adapt to their environment. The book considers the evidence of the pessimistic struggles and optimistic pleasures in both Darwin's texts and correspondence and discusses whether they are Romantic or not. It also examines Darwin's quest to understand the 'struggle' in Nature as well as his struggles within his own scientific, intellectual and personal development and whether these can be regarded as Romantic.

There are many metaphors in Darwin's works that can be regarded as Romantic. David Kohn, for example, traces the origins of Darwin's entangled bank metaphor back to the impenetrable thickets of Milton's *Paradise Lost* experienced by Darwin first hand as a struggle towards the sublime when going through the forests of Tierra del Fuego and Brazil (Kohn, 1996, p. 26). Yet the 'bank' element of the 'entangled bank' can be interpreted as a place of quiet beauty, a garden of Eden 'from which all human's spring' (Kohn, 1996, p. 29) despite being a place reflecting 'the war of nature', 'famine and death' (Darwin, 1985, p. 459). Kohn traces Darwin's 'wedges' metaphor back to Darwin's Malthusian *Transmutation Notebook D* in which the wedges form gaps by 'thrusting out weaker ones' (Barrett,

---

13 The 'entangled bank' represents the diversity of Nature with its varieties of plants, birds, insects and worms all dependant on each other (see Darwin, 1985, pp. 459).

2008, p. 375). This metaphor 'captures the elevated experience of insight, while in part it conveys the scientific content of that insight. These two dimensions of the wedging metaphor – elevated experience and scientific content – fuse' (Kohn, 1996, p. 37). According to Kohn, the aesthetic practice of incorporating such metaphors as the 'entangled bank' and the 'wedges' in his texts, showed that 'Darwin's *Beagle* enterprise was conducted according to one of the characteristic norms of Romantic science, namely: Darwin self-consciously incorporated affect and imagination into his early science' (Kohn, 1996, p. 15). The metaphors are also examples of Darwin combining the personal with the objective. Darwin first 'contemplates nature and reflects on his own assumptions' but then invites the reader to join with him in changing his assumptions due to his discoveries. So Darwin starts off observing Nature objectively but then combines these with his subjective imaginative interpretations of Nature and invites his reader to do the same (Kohn, 1996, p. 40). Kohn sees this as reminiscent of the landscape painting convention in which 'a human figure is put in a corner to overlook a picturesque or sublime scene'. In Darwin's texts, he is not just an objective observer but is also a participant observer contributing his own thoughts and emotions to the scene. Thus 'the drama of Darwin's psychic life is implicated in the metaphor' (Kohn, 1996, p. 42) and can be regarded as Romantic.

Darwin's contemplation and reflection of Nature, an essential part of his Romantic imagination, can be seen in the concept of 'rastro'. Darwin introduces this concept in his *Voyage*[14] when referring to the native Indians in search of the wild Indians who had committed murder and who had left a track ('rastro') leading into the Pampas (Levine, 2011, p. 58). One 'glance at the rastro' enabled the native Indians to work out the history of the horses' journey:

> One glance at the rastro tells these people a whole history. Supposing they examine the track of 1,000 horses, they will soon guess by seeing how many have cantered the number of men; by the depth of the other impressions, whether any horses were loaded with cargoes; by the irregularity of the footsteps, how far tired; by the manner

14 In August 1833 while in Bahia Blanca.

in which the food has been cooked, whether the pursued travelled in haste; by the general appearance, how long it has been since they passed. They consider a rastro of ten days or a fortnight, quite recent enough to be hunted out. (Darwin, 1989, p. 110; see also Keynes, 2009, pp. 214–15)

In my interpretation of Darwin's use of 'rastro', the idea of 'track' or 'trace' is only part of the meaning. 'Rastro' can also be seen to represent layers of fossils indicating different periods or traces of history which can also include biographical traces of Darwin's own history, his published and unpublished works and his correspondence. This can also include traces of the influence of others on his ideas such as his grandfather Erasmus Darwin, Alexander von Humboldt and Goethe. The expression of Darwin's Romantic language flowing through his works can be seen to be an imaginative collection of ideas in poetry and prose taking in the past, present and future. In this respect, the 'rastro' can be seen as a set of unfolding ideas making up a narrative of the development of Man's Mind and the mind of Darwin. The 'rastro' or 'rastros' are Romantic because they relate the objective forms of Nature to its subjective reading; that is, they represent the oneness of nature in which everything is related to everything else.

Darwin's Romantic imagination can be interpreted as a poetic 'rastro' narrative. This is best illustrated with reference to Percy Bysshe Shelley's essay 'A Defence of Poetry' (Shelley, 2008).[15] For Shelley, poetry is 'the expression of the imagination' (Shelley, 2008, p. 40), in which the thoughts of the 'imagination' are on a higher plane to that of the thoughts of 'reason'. By this he means that 'imagination' is the awareness of more than what is currently experienced and it is this which creates human values. An example of this imaginative interpretation of Nature can be seen in Humboldt's method of viewing Nature (the 'Humboldtian Method'), in which the description of a geographical area is more than its scientific data and takes into account the aesthetic impression the landscape makes on the observer that make the area unique. As with Darwin, language is the creator of Man: 'In the infancy of society every author is necessarily a poet, because

---

15 Published posthumously in 1840 but composed in 1821 a year before Shelley's death, when Darwin was still a twelve-year-old.

language itself is poetry' (Shelley, 2008, p. 42), thus 'poetry is connate with the origins of man' (Shelley, 2008, p. 40). Here the 'rastro' is more than just a collection of evidence (or 'enumeration of qualities already known' making up 'reason')'; it is an imaginative collection of ideas taking in the past, present and future. Here the writer or poet unites the characters of a legislator and a prophet as 'he beholds the future in the present, and his thoughts are the germs of the flower and the fruit of latest time' (here Shelley uses the term 'prophet' to capture the sense that 'a poet participates in the eternal, the infinite, and the one') (Shelley, 2008, p. 43). Here the 'rastro' and the 'imagination' are one in uniting all that has been, with all that will be, in terms of fossils, transmutation of species, their histories and their origins. This concept of the 'rastro' and the 'imagination' is poetically expressed by Shelley's own 'enlarged imagination':

> [Poetry] is as it were the interpretation of a diviner nature through our own; but its footsteps are like those of a wind over the sea, which the coming calm erases, and whose traces remain only, as on the wrinkled sand which paves it. These and corresponding conditions of being are experienced principally by those of the most delicate sensibility and the most enlarged imagination. (Shelley, 2008, p. 63)

The objects that have made the footsteps have disappeared, but the evidence or traces of their past existences remain in the form of the 'wrinkled sand' (or footprints, or in the case of Darwin, 'rastros').

This poetic 'rastro' narrative is what Kohn calls a 'romantic aesthetic'. This is also a form of 'enlarged imagination' similar to the Humboldtian Method of combining objective scientific data with the subjective emotional feelings of the observer. Kohn sees Darwin's Romantic aesthetic as a form of contemplation in which Darwin first shares his observations of Nature objectively with his reader, then reveals his response to it emotionally and then finally reveals 'his identification with his explanatory theory of nature' (Kohn, 1996, p. 14). This is a form of conversation between Darwin's self and his imagination and between his self and his reader. Darwin refers to this conversation as 'double consciousness' in his *Notebook N* (Barrett, 2008, p. 593).This is a direct reference to William Wordsworth's use of the expression in his poem *The Excursion* which Darwin had read several times (Beer, 2010, p. 8). This excursive power of the mind enabled Darwin to move between science and imagination (his 'castles in the air') (Barrett, 2008, p. 527), and between the humble physical beginnings of Man and Man's moral being (Beer, 2010, p. 8).

This importance of Mind in Darwin's writing is highlighted by his 'rastro' method. This underlines the significance of contemplation and reflection which are important aspects of morality in Man. Darwin shows how morality in Man is developed from instincts and how this leads to the emotion of sympathy. Sympathy is an essential part of Darwin's own thought processes in feeling that all creatures share a common consciousness. This not only enables Darwin to feel a common empathy with animals but enables him to get into their 'minds', thus understanding sexual selection for example. In examining Darwin's 'rastros', it would be impossible not to include the traces of his grandfather Erasmus Darwin's influences in his life and works, although Charles Darwin was adamant that he had not been influenced by him in any way. The book presents evidence suggesting that the works of Erasmus Darwin do reflect a similar philosophy of Nature to Humboldt and Darwin's 'one reality' Nature with its 'web of affinities'.

This book examines the principal works of Charles Darwin, viz. *The Voyage of the Beagle* (1839), *On the Origin of Species* (1859),[16] *The Descent of Man* (1871) and his *Notebooks (1836–1844)*. Particular reference is made to the Romantic concepts employed by Alexander von Humboldt in his works as Darwin took Humboldt's *Personal Narrative* (1807–1834) with him on the voyage of the *Beagle* (1831–1836) and frequently praised Humboldt in his *Voyage of the Beagle* for his descriptions of Nature. To understand the Humboldtian concepts of Nature, one needs to understand the 'Humboldtian Method' of naturalist research. It demonstrates that including the observer's aesthetic experience of Nature alongside scientific observations of measurement, enables 'art' and science to complement each other. Together they enable the naturalist, poet and artist (and many

---

16   The edition referred to throughout the book, apart from one exception, is the 1859 first edition (Darwin, 1985). The reason for this is that Darwin weakened his argument in later editions by making concessions to his critics. The earliest version also brings the reader closer to Darwin's thought processes during the time the theory was first conceived (Burrow, 1985, p. 49).

naturalists in the eighteenth and nineteenth centuries were all three) to trace the history of creation and diversity, whether this be in the plants or the rocks, and what makes particular regions distinct. Darwin learnt from Humboldt that making comparisons between species and between regions gave a greater insight into diversity and the interrelationships between the underlying laws responsible for it. In comparing Darwin's concepts to Humboldt's, Darwin's theory of natural selection can be seen to sum up Humboldt's organic interrelated Nature as everything is related to everything else, and the sum total of Nature is the sum total of its own creation, its underlying forces and laws, as it is its own creator. Like the 'entangled bank' in Nature, all the relationships work together to produce change in order to adapt and survive. Nothing stands still in time.

Darwin's aesthetic imagination was not only honed by Humboldt, but also by Dugald Stewart's concept of the sublime. In his notebooks, Darwin refers to Stewart's use of the sublime, in which it is defined as an active power of soaring upwards, a flight of the imagination, as if created by a supernatural heavenly agent raising those thoughts towards heaven (Stewart, 1829, p. 270). For Darwin this imaginative flight is often 'accompanied with terror and wonderment' (Barrett, 2008, p. 605). It is this imaginative ability which enables Darwin to sympathize with, for example, the common beetle or barnacle, by identifying his subjective self with objective Nature, making them one. This sympathetic feeling of the connection between the subjective and the objective that makes up the wholeness of Nature, unveils the 'vastness of Eternity' creating the impression of a Deity. This is Darwin's feeling of an 'inward glorying' (a sense of inward greatness) or sublime ecstasy (Barrett, 2008, p. 605), and is the kind of experience Darwin would have felt when germinating seeds from bird droppings, from which new life comes from what appears to be 'sordid' and 'messy' (Levine, 2008, p. 168). This 'inward glorying' or 'enlarged imagination' is also necessary to help Darwin interpret the 'rastros' and write the narrative.

To understand these notions within the context of Victorian society, it is important to understand the significance of Prince Albert's contribution. Albert came to Britain with many German ideas helping to lay the foundation stones for intellectual change. Prince Albert had a considerable impact on Victorian society in championing the ideals of progress in the arts and sciences and making such developments acceptable both politically and socially. This acceptance of change and development within science and technology made it easier for the works of such naturalists as Darwin to gain greater acceptance, particularly when it came to the reception of the idea of the transmutation of species. Early in 1845, for example, Prince Albert read *Vestiges*[17] to his young queen and this led to other families reading and discussing the work alongside other publications on Nature culminating in the publication of Darwin's *Origin* in 1859. Albert had a great interest in science going back to his childhood in Coburg where he was taught chemistry, natural philosophy, and natural history, and he hoped to bring this interest in science to the British people (Secord, 2000, pp. 168–9). In 1847 'Albert commissioned research into the curriculum' at Cambridge University and discovered that apart from mathematics it was 'based entirely on knowledge of the Classics and the Scriptures' (Heffer, 2013, p. 447). After a lot of initial opposition, Albert, with support from the Vice Chancellor, persuaded the Senate in 1848 to approve a degree course in natural sciences, history and moral sciences (Heffer, 2013, p. 447). Albert was also the driving force behind the Great Exhibition of 1851 which 'showcased British innovation, engineering and design [...] and was one of the turning points of the Victorian age' (Heffer, 2013, p. 285). It 'was designed to stimulate a nationwide interest in science and the arts, the former reflecting Albert's very German idea of the importance of technological advancement' leading to the new order of social and economic change (Heffer, 2013, p. 286). Ironically those institutions that wanted to be a part of this new 'national focus of cultural life' helped underpin 'the civilizing goal of the pursuit of perfection' (Heffer, 2013, p. 286). Prince Albert, therefore, indirectly helped pave the way for the reception of Darwin's *Origin*.

---

17   Robert Chambers work, published anonymously in 1844, detailing the traces of change in Nature over thousands of years. This is akin to Darwin's 'For myself, I venture confidently to look back thousands on thousands of generations' (Darwin, 1985, p. 201), in which he sees 'incipient species', or variations, as the link between species and new species.

The chronological order of the book attempts to mirror Darwin's own methods and creative insights as expressed through the influence of the 'Humboldtian Method', Goethe's 'Genetic Method' and Darwin's 'rastro' method. These methods enable the observer to gain a deep understanding of the secrets of Nature through a to-ing and fro-ing in place and time. This could be anything from making links between past progenitors and present Man through the discovery of fossils or shells above sea level and present rudimentary forms once of use to our forebears such as wisdom teeth; or this could be the mind's ability to move between the particular and universal archetypes, as expressed by Goethe's 'leaf' archetype which captures the underlying form responsible for the metamorphosis of the seed, the root, the leaf and the flower of a plant. My book intentionally starts with Darwin's *Origin* as this is his central work and his *The Descent of Man* is an extension of this. But Darwin's key theory of natural selection was developed from his data gathering and experiences during his five-year voyage on the *Beagle*. The older Darwin, both in terms of the man and his works, cannot therefore be understood without reference to his *Voyage of the Beagle*. Reading the *Voyage* on its own without any reference to the *Origin*, it can be seen almost as a naturalist's adventure story or travelogue. But when the reader goes backwards and forwards between the *Origin* and the *Voyage*, the embryonic theory of natural selection can clearly be seen in the *Voyage*. It is therefore the intention of the book to move freely between the old and the young Darwin, moving backwards and forwards in time in order to fully understand his thought processes. As with Humboldt and Goethe before him, this movement between insights in time and place should be able to create an interrelated organic overview of Darwin's theory of Nature revealing all his various shades of Romanticism like the feathers of a peacock.

From an initial reading, as one would expect, the language of the young Darwin is more youthful, more spontaneous and more poetic and can be seen to be very similar in form to Humboldt's *Personal Narrative*, and Darwin was chided by his sister Caroline for this.[18] All his experiences

---

18   In a letter to Darwin from his sister Caroline on 28 October 1833 regarding the journals sent to her from the *Beagle*: 'As to your style. I thought […] that you had,

are new and he is overcome by wonder and this is expressed in his use of language. The later Darwin in the *Descent* is written in a more scientific, and perhaps more Victorian, vein. He is in his sixties and is an establishment figure. He has more to lose in terms of his reputation and therefore has to be more careful in the way he expresses himself. From an initial reading of the prose there are still hints of Romanticism, it could be argued, but they are more difficult to find, hidden under a heavy text of detailed evidence (always, almost obsessively, providing perhaps more detail than required due to a fear of others trying to prove him wrong). But for the reader it suddenly becomes all worthwhile when one comes across Darwin's descriptions and illustrations of the peacock and the Argus pheasant in which their respective rich histories of sexual development can be seen in the ocelli (or 'eyes') on their feathers. Just like a fossilized imprint (or 'rastro') of their development frozen in time, they demonstrate 'that a gradation is possible' in which spots are converted into 'elliptic ornaments for the purpose of sexual selection' (Darwin, 2004, pp. 495–6).

The texts of the earlier Darwin and the later Darwin are examined using *The Voyage of the Beagle* and *The Descent of Man* along with their corresponding notebooks. This is used as an opportunity to compare and contrast the language used in each of the texts to highlight the differences in the expression of Romantic thought through his language. In addition to this, Darwin the Victorian is compared to Darwin the Romantic to determine whether there are any conflicts or contradictions between the two. Are there pressures on the establishment Victorian figurehead to rein in his Romanticism and therefore his evolutionary theories? There certainly seem to be some ideological differences between the two Darwins; for example the older Darwin saw evolution as an example of improvement through all the stages of humanity such that the savage was at a lower stage of improvement to that of the Victorian gentleman, whereas the younger Darwin seemed to think of all humanity being on the same level. The older Darwin, perhaps due to the civil unrest at the time both in America and

---

probably from reading so much of Humboldt, got his phraseology [...] instead of your own simple straight forward & far more agreeable style.' (Burkhardt, 2008, p. 236).

England, ironically displayed a degree of racism himself regarding Irish workers in New York as lesser beings: 'What devils the low Irish have proved themselves in New York' (cited by Desmond and Moore, 2004, p. xlv). Darwin even thought women were at a lower stage of development[19] than men: ['Intuition', 'imitation' and 'rapid perception' in women characterized] 'the lower races, and therefore ... a past and lower state of civilization' (cited by Desmond and Moore, 2004, p. xlviii).

The text of Darwin's *Descent*, alongside his letters and notes, is examined to determine whether these changes and any others constitute a shift away from Romanticism. When Darwin was on the *Beagle* he was horrified by the sight of slaves, and although equally horrified by the primitive way of living of some of the native tribes he met on his travels, he regarded both slaves and primitive tribes as human with equal rights as they ultimately came from the same progenitor and were not a separate species. Darwin's extension of his theory of natural selection to include the addition of his theory of sexual selection is critical in this regard. Darwin noticed that

---

19 Darwin's *Origin* inspired his cousin, Francis Galton to research human intelligence which he believed to be hereditary. He coined the term 'eugenics', believing that marriage between families of high rank should be encouraged in order to produce intelligent offspring, and that geniuses could be created by selective breeding just like the selected breeds created by pigeon fanciers as outlined in Darwin's *Origin* (Galton's theory was published in his book *Hereditary Genius* 1869) (Desmond and Moore, 2004, pp. xlvi–lvii). Darwin's *Descent* prescribed a form of this eugenics by advocating that 'both sexes ought to refrain from marriage if in any marked degree inferior in body or mind'; 'all ought to refrain from marriage who cannot avoid abject poverty for their children'; and following Malthus, 'our natural rate of increase, though leading to many and obvious evils, must not be greatly diminished by any means' (cited by Desmond and Moore, 2004, p. liii). Darwin's son, George, also supported these views. This is an example of Social Darwinism that got a bad name after being taken up by the Nazis and used against the Jews in the Second World War. The analysis of post-Darwin Social Darwinism is beyond the scope of this book. However, it is important when researching Darwin to make a distinction between his views and his theories and not to make a category error in conflating an 'is' with an 'ought'; that is, describing a state of affairs (for example, the law of natural selection) is different from prescribing a state of affairs (for example, 'only the well-to-do should breed').

humans' skin colour, shape and form did not directly correspond to region or habitat as those from the same climatic region did not necessarily have the same skin colour. Darwin attributed this to sexual selection. According to Darwin, the concept of beauty varied regionally and males and females selected each other within their respective regions according to what was held to be beautiful in their regions. Darwin argued that the same applied to the animal kingdom. The importance of this argument for Darwin is that it further strengthens his view that humans of whatever colour can be traced back to the same progenitor and that although their differences make them different races, they are all varieties of the same species. In terms of the argument for defining Darwin's concepts as Romantic, this may seem to be a digression. But the argument for sexual selection can be seen as another example of the 'web of affinities' in that the whole of existence, whether this be plant life, animal life or human life, is all related, and that the past histories of all life are also all connected to the present. As such sexual selection can be seen as an essential spoke in the wheel of Organic Nature.

Chapter 1 examines Humboldt's unity of Nature in which everything is related to everything else and how this is developed by Darwin into his own version of the 'web of affinities', in which Nature's 'One Reality' creates itself through its own laws without the need for an external Deity. It also examines Humboldt's method of including his impressions of Nature in his observations which played a significant part in the development of Darwin's Romantic imagination. The chapter also examines the influence of William Wordsworth in the development of Darwin's use of 'double consciousness'. Chapter 2 continues to trace the influence of Humboldt in the development of Darwin's Romantic imagination showing how Humboldt's aesthetic appreciation of Nature helped Darwin 'see' the laws of Nature at work. Through their common abhorrence towards slavery, this chapter also underlines their view that as Man comes from the same progenitor, no race should be treated as inferior to another. Chapter 3 traces the roots of Humboldt's and Darwin's Romanticism, further examining Humboldt's aesthetic method and how this sparked Darwin's own imagination into seeing natural selection in Nature. Goethe's 'Genetic Method' is introduced in this chapter to highlight Darwin's imaginative ability of moving backwards and forwards in time in order to understand

Nature's development. Although not Romantic, Malthus and Townsend are included to show their influence on Darwin in terms of the concepts of struggle and self-improvement used in his theory. Chapter 4 argues that Man's view of Nature is not just made up of a set of empirical perceptions but also consists of a set of mental perceptions reflecting on them. Moral reflection is one of the highest forms of mental reflection and Chapter 5 shows how this, like aesthetic reflection, can be regarded as Romantic as it developed from material Nature over aeons of time. Chapter 6 provides a sociological appraisal of Darwin's own moral values within the context of the Victorian era enabling the contextualization of the Romantic concepts such as 'Nature' and 'Mind' examined in previous chapters. Chapter 7 traces the development of Darwin's Romantic concept of Mind from the *Voyage* and his notebooks to the *Origin* and the *Descent* and how his Romantic imagination evolved from his material experiences of Nature. Charles Darwin denied that his grandfather Erasmus Darwin had influenced him, but Chapter 8's analysis of Erasmus Darwin's poetry suggests that there is evidence to show that Darwin's Romantic materialism was influenced by his grandfather.

In offering a conclusion to the book's research, the last chapter is devoted to an analysis of Ruth Padel's biographical poem of her great-great grandfather, Charles Darwin (Padel, 2010). Her poem is cross-referenced with Coleridge's *The Rime of the Ancient Mariner* and Percy Bysshe Shelley's 'A Defence of Poetry' in order to underline the nature of Darwin's Romantic 'poetry'.

Through her poetry she is able to reflect the Romantic nature of his one reality Nature through its interrelatedness, seen in both his personal life and in his theories of Nature. Padel's sensitive poetry inspires the reader to enter the mind and body of Darwin in the same way as Darwin was able to imagine entering the minds and bodies of worms, birds, animals and other beings, and in so doing to experience the sublime beauty of Nature. Ruth Padel's poetry not only draws out the Romanticism of Charles Darwin but also enables the reader to understand and experience the Romanticism of Nature that is, in effect, the poetry of Nature: from Death comes Life, from Matter comes Mind.

CHAPTER I

Organic and One Reality Nature in Humboldt and Darwin

This chapter will examine the influence of Humboldt's 'Organic Nature' in the development of Darwin's 'web of affinities' to help identify the roots of Darwin's Romanticism. In order to understand Humboldt's organic view of Nature,[1] it is useful to identify which thinkers and their ideas had helped Humboldt develop this view. It is also helpful to provide some background to exploration at the time in order to understand the context of naturalists' ideas.

Humboldt had read Immanuel Kant[2] whose own organic way of thinking had been sparked off through his dissatisfaction with Carl Linnaeus' classification system of the natural world;[3] Kant saw this system as incapable of conveying the idea of the whole by not placing enough emphasis on the unifying processes and the interrelationships between the parts and too much emphasis on their external structures (McCrory, 2010, p. 122). What was missing was a description of the interrelated unifying processes

---

1   The 'N' in 'Nature' is used throughout the book in upper case. This is to identify the complexity of its use as it can refer to empirical instances before the observers' eyes or abstract concepts. It may refer to one or the other or may refer to both at the same time. The context of the word's use in this book should make its meaning clear. Lower case 'n' will only be used where its use occurs in quotes from other authors.
2   Humboldt refers to Kant's *On the Theory and Structure of the Heavens* (1755) in which he says that the planets, stars and earth make up the organic whole that is Nature (Humboldt, 1997, p. 50 and p. 65).
3   However, building on the Linnaean system, modern taxonomy follows the principle that taxonomic groups should be based on evolutionary relationships. 'Modern classifications are viewed as evolutionary hypothesis about the relationship of organisms' (Davidson, 2012, unnumbered page).

that could not be directly seen. Humboldt filled this gap by describing the character of each geographical region of land through these interrelationships, such as the hills, the plants and the colour of the sky[4] (Nicolsen, 1995, p. xiii). Humboldt had also been influenced by Johann Forster[5] and his son George Forster, both of whom had been with Captain James Cook on the second of his three voyages of discovery[6] to the Pacific. Humboldt travelled with George Forster around Europe and Forster's publication of their journey in 1790, *Ansichten vom Niederrhein von Brabant, Flandern, Holland, England und Frankreich*, 'was a demonstration that scientific inquiry need not be cold and unresponsive to the beauties of Nature. It could embrace and celebrate the earth in the act of studying it' (Nicolsen, 1995, p. xv). The word *Ansichten* refers to both a visual view and a personal view or opinion. This is the approach taken up by Humboldt's empiricism that encapsulates both the subjective and the scientific, as well as his ecological approach to Nature (Sachs, 2007, p. 55). This was a turning point in the way Nature should be viewed. No longer were plants and features of the earth to be seen in isolation. For Humboldt, a history of the earth should not just concern itself with geological phenomena but should also concern itself with the diversity of biological phenomena. For Humboldt, voyages were not only about exploration and the adventure of discovering new continents and peoples of the world, but also about understanding the diversity that was underlying the unity of Nature (Nicolsen, 1995, p. xvii).This diversity could be understood through Humboldt's method of

---

4 According to Nicolsen, 'In his *Florae Fribergensis specimen* (1817), Humboldt followed Kant in distinguishing between a true history of nature and a mere description of natural objects such as had been provided by the older Linnaean system' (Nicolsen, 1995, p. xvii).

5 Forster senior published his account of the voyage with Captain Cook as *Observations Made during a Voyage round the World* in 1778. Forster influenced Humboldt through his experimental methods and his theory of the environmental influences of climate, winds etc. on vegetation producing the diversity of forms. For both Forster and Humboldt, the study of vegetation was a key part in the understanding of Nature as this impacted most on Man (Nicolsen, 1995, pp. xiii–xiv).

6 His second voyage was between 1772 and 1775 in which he circumnavigated the globe, sailing close to the Antarctic in the east and west.

blending science with aesthetic sensitivity (the 'Humboldtian Method'); this was a move away from an analysis of individual specimens (as per Linnaeus) to an analysis of how the individuals and species interacted with each other and their environment, including the sensitivities of the observer and how they impacted on him/her.[7]

In Humboldt's time, exploration was still very much part of colonial expansion.[8] As a German, Humboldt was fortunate, through his contacts, to get permission from the Spanish king to explore Spanish America as this was under Spain's control and was out of bounds to foreigners. As a mining geologist, he was seen as an asset to further developing their colonies. At the time of his visit to America (1799–1804), Humboldt had all the latest scientific equipment and was better equipped than any previous explorers, enabling him to take detailed readings of the terrain, climate, plant life, animal life and geography. Previously most explorers had limited their travels to the coast for mapping and trade purposes whereas Humboldt was one of the first to go into the interior.

In Humboldt's *Personal Narrative*[9] the reader familiar with Darwin's *Origin of Species* can see some of the similar questions being posed and a hint at some of the similar answers being proposed. These are questions dealing with the concepts of organic forms, their connections, relationships,[10]

---

7   This is the very foundation of modern ecology.
8   Captain Cook's voyage on the *Endeavour* was primarily arranged by the Royal Navy to measure the transit of Venus in 1769 in order to calculate the distance of the Earth from the Sun, having been beaten by the French in their last voyage in 1761. However, Cook was also secretly instructed by the government that the objective was also to acquire new territories showing that such voyages were also concerned with colonial expansion (Fara, 2003, pp. 74–5 and p. 77).
9   It took Humboldt thirty years and thirty volumes to publish his American expedition. The *Personal Narrative* part of his publication, which covered the popularization of his actual journey, appeared in volumes 28 to 30 and was published in French as *Relation historique du voyage aux régions équinoxiales du nouveau continent* in 1814, 1819 and 1825 (Sachs, 2007, p. 46; Wilson, 1995, pp. lv–lvi; Botting, 1973, pp. 2002–7 and McCrory, 2010, pp. 54–5 and p. 118).
10 The term 'relations' is probably the most common term in Humboldt's *Personal Narrative* (Sachs, 2007, p. 49). In the same way as Humboldt establishes the 'relations' between rocks, wind and water, he also establishes 'relations' between his

similarities and interrelatedness forming the unity of life. An analysis of Humboldt's discussion of migration and the development of identical species in different areas reveals some of Darwin's embryonic thoughts on his theory of natural selection:

> Even when nature does not produce the same species in analogous climates [...], we still noticed a striking resemblance of appearance and physiognomy in the vegetation of the most distant countries. This phenomenon is one of the most curious in the history of organic forms. I say history, for reason cannot stop man forming hypotheses on the origin of things; he will always puzzle himself with insoluble problems relating to the distribution of beings. (Humboldt, 1995, p. 138)

This groping towards a position later taken up by Darwin can be seen in the words 'striking resemblance of appearance [...] in the vegetation of the most distant countries'. The importance of this is that Humboldt is breaking with the view that the Creator, as a first cause, created these similarities. Humboldt is here suggesting that change might be responsible as a result of interactions between species and their environment, that there might be a connection between these resembling species from different countries and that they might have a common heritage. The key word that hints at change is 'history': the notion that 'organic forms' may have a 'history' suggests that over time they may change and that time may enable further change to take place in different places (that is, the hint that not all species necessarily migrate as they now appear, but migrate and then change as they adapt to their new environment). This enables him, like Darwin after him, to form 'hypotheses on the origin of things' and to consider the 'problems relating to the distribution of beings'. Here the concept of 'distribution' relates to all the other factors concerning the modification of species and how they move to different regions, and, at the same time, how they resemble each other.

'scientific tangents' such as his descriptions of volcanoes and earthquakes, and these 'relations' create the same unity in his book as exists in Nature (Sachs, 2007, p. 49). As expressed by Humboldt, 'I begin by describing the phenomena as they appeared to me, then I consider their individual relations to the whole' (Humboldt, 1995, p. 12).

Humboldt is not convinced that migration is responsible for similar species appearing at the same altitudes; he believes that there is something else as yet unknown responsible for this, but something to do with the species themselves and how they interact with their environment:

> Neighbouring species are often found at enormous distances from each other, in low regions of a temperate zone, and on mountains on the equator. And, as we found on La Silla at Caracas, they are not the European genera that have colonized mountains of the torrid zone, but genera of the same tribe, which have taken their place and are hard to distinguish. (Humboldt, 1995, p. 139)

These observations are hints at ideas later taken up by Darwin; for example, species of the same 'tribe' (or genera) not necessarily having migrated as they are but having taken up different habitats and adapted to them, resulting in changed appearances, yet, through their similarities, still showing a relationship to their original forms. Humboldt had not arrived at this fully formed idea, but his thoughts can be seen as moving in this direction.

In abandoning the hypothesis that migration is responsible for the distribution of species, Humboldt also abandons the view that identical temperatures are responsible for a certain kind of vegetation growing in specific regions:

> It is tacitly assumed that under identical temperatures a certain kind of vegetation must grow. This is not strictly true. The pines of Mexico are absent in the Peruvian Andes. The Caracas la Silla is not covered with the same oaks that flourish in new Granada at the same height. Identity of form suggests an analogy of climate, but in similar climates the species may be very diversified. (Humboldt, 1995, p. 139)

There is therefore something else responsible for this diversification. Here Humboldt is doing some of the scientific spadework for Darwin enabling him to distil his theory of natural selection, but also (as discussed later) helping him develop the 'Humboldtian Method' of seeing the world differently from the way naturalists had viewed it before. It is Humboldt's notion of the organic unity of life's relationships (between species and species, and species and their environment) that makes this possible.

The concept of relationships in Nature and how everything has an effect on everything else is an important one in Humboldt's works and has an important influence on Darwin in the development of his own theory of Nature. The cause and effect of relationships in Nature leads to change or transmutations in Nature. Humboldt shows how the geographical distribution of plants is due to the relationship between plants, climate, terrain and altitude (Humboldt, 1995, pp. 95–6). The relationships of the features of a specific area make that area's characteristics unique and therefore easily identifiable and distinct from other areas. Humboldt underlines the importance of this when comparing the Orinoco to other river beds:

> These differences do not depend solely on the width or speed of the current; they derive from a combination of relations easier to grasp on the spot than to define precisely. In the same way, the shape of the waves, the colour of the water, the kind of sky and clouds, all help a navigator guess whether he is in the Atlantic, the Mediterranean or in the equinoctial part of the Pacific. (Humboldt, 1995, p. 186)

This way of seeing was important to Darwin in being able to relate the histories of rock strata and their fossils with the development of the surrounding plant and animal life. It enabled him to see the histories and developments of species in space and time rather than fixed unchanging entities frozen in present time. In Humboldt's study of earthquakes, he goes beyond the collection of facts stressing the importance of comparing them in order to go beyond what is variable. 'We get to know the features of each region better the more we indicate its varying characteristics by comparing it with others' (Humboldt, 1995, pp. 34–5). This is unity in diversity. Comparisons can lead to the discovery of the laws that make things variable. This again is a hint of what is to be developed by Darwin:

> When we cannot hope to guess the causes of natural phenomena, we ought at least to try to discover their laws and, by comparing numerous facts, distinguish what is permanent and constant from what is variable and accidental. (Humboldt, 1995, p. 59)

Humboldt also taught Darwin that making comparisons through firsthand experience of the forces of Nature, despite its dangers, was a better way to learning about and understanding the laws of Nature. Humboldt's direct experience of earthquakes, literally breathing in the sulphurous gases around him and feeling the ground move, enabled him to make the connection between earthquakes and volcanoes:

> Thus sitting in the interior of a burning crater near those hillocks formed by scoriae and ashes, we feel the ground move several seconds before each eruption takes place. We observed this phenomenon at Vesuvius in 1805 [and] in 1802 on the brink of the immense crater of Pichincha [...]. Everything in earthquakes seems to indicate the action of elastic fluids seeking an outlet to spread into the atmosphere. (Humboldt, 1995, p. 64)

This made Humboldt break with the view that volcanoes were merely the result of burning matter beneath the surface. It made him realize that there were much stronger hidden forces at play. As a 'natural historian', Humboldt realizes that establishing relationships between phenomena helps the naturalist understand Nature's history, and thereby its inner secrets, its inner laws:

> As a historian of nature, the traveller should note down the moment when great natural calamities happen, and investigate the causes and relations, and establish fixed points in the rapid course of time, in the transformations that succeed each other ceaselessly so that he can compare them with previous catastrophes. (Humboldt, 1995, p. 142)

In examining these 'transformations that succeed each other ceaselessly', Humboldt is informing his reader of his quest to discover the permanent laws of Nature that create this 'process of constant change' (McCrory, 2010, p. 81).

Darwin also experienced earthquakes on his voyage and his reading of Humboldt's experiences must have been an aid to his own understanding of such phenomena. We know from Darwin's letters that Adam Sedgwick[11] arranged for him to have a copy of Humboldt's *Personal Narrative* with him on the *Beagle*: 'Humboldts [sic] personal narrative you will of course get – He will at least show the right spirit with wh. a [sic] man should set to work' (Burkhardt, 2008, p. 47). In a letter to his Cambridge mentor John Stevens

---

11   Although Darwin was not formally studying geology at Cambridge, he developed his knowledge of geology through his friendship with Sedgwick (Professor of Geology) and through going on field trips with him.

Henslow[12] from the *Beagle* in 1832, Darwin refers to his Humboldtian experience of the unity of Nature on an expedition in South America to Rio Macaò:

> Here I first saw a tropical forest in all its sublime grandeur. – Nothing, but the reality can give any idea, how wonderful, how magnificent the scene is. – If I was to specify any one thing I should give the preemenence [sic] to the host of parasitical plants [...]. I never experienced such intense delight. – I formerly admired Humboldt, now I almost adore him; he alone gives any notion, of the feelings which are raised in the mind on first entering the Tropics.[13] (Burkhardt, 2008, p. 128)

Darwin had read descriptions of the Tropics in Humboldt which inspired him to travel. Everything was so completely different from anything he had seen in Britain. To see whole gardens of plants growing on other plants was an incredible sight for a naturalist, especially at that time when he would only have had access to books and drawings (and as Humboldt had observed, many explorers copied from each other creating a distorted view of reality). This first fresh experience of the Tropics gained first-hand opened up his mind to the unfathomable possibilities of what Nature could be and how it got to that position. As all things seemed possible it opened up his mind to the possibility of all things biologically and botanically. This openness is crucial to an understanding of how Darwin was able to collect data, analyse the data comparatively in relation to other data rather than in isolation, and to be open to all possible interpretations which others might have missed at the time as well as the development of his new insight of inter-relatedness.

Humboldt (along with other important philosophers and natural historians of the time such as Schelling and Goethe), had read Carl Kielmeyer's published lecture 'On the relationships of the organic forces' (1793)[14] and

---

12 Henslow received many specimens from Darwin while on his *Beagle* voyage (Burkhardt, 2008, p. 420).

13 As referred to in the passage of Humboldt's experience of the Tropics quoted elsewhere in this chapter.

14 Published in German as *Ueber die Verhältnisse der organischen Kräfte unter einander in der Reithe der verschiedenen Organisationen, die Gesetze und Folgen dieser Verhältnisse* (Stuttgart, 1793; reprinted Tübingen, 1894).

had been influenced by his argument that organs in an organism should be seen as cause and effect of each other.[15] For example, the heart and lungs work together to support each other's existence as do other organs in the human body. This relationship could be seen as not just being between organs in individual organisms but also as being between members of the same species, for example, between male and female members, and between members of different species, for example, cats killing mice that damage bumble bee nests thereby improving clover and honey production through pollination, (Darwin, 1985, p. 125) and bumblebees nesting in disused rodent nests.[16] This organic concept of Nature viewed as both product and creator of itself (following Spinoza's theory of Nature) was therefore not purely mechanistic (Richards, 2002, p. 116; p. 241; p. 539). Darwin sums this up in the mutual relationship between the bee and the flower:

> Thus I can understand how a flower and a bee might slowly become, either simultaneously or one after the other, modified and adapted in the most perfect manner to each other, by the continued preservation of individuals presenting mutual and slightly favourable deviations of structure. (Darwin, 1985, p. 142)

Humboldt also saw this mutual relationship between species as being under threat when Man failed to realize how his own behaviour towards Nature could produce a negative impact. Just as Thomas Malthus[17] saw overpopulation as having a negative effect on the amount of food available, Humboldt could demonstrate how deforestation through the felling of trees reduces the supply of fuel and creates a scarcity of water, and as such he could be seen as an early ecologist (coming from his notion of interdependent unity):

> Thus, the clearing of forests, the absence of permanent springs, and torrents are three closely connected phenomena. Countries in different hemispheres like Lombardy

---

15 According to Humboldt's book dedication (Humboldt, 1806), this essay made Kielmeyer 'the first physiologist of Germany'. Goethe had met Kielmeyer at Tübingen in September 1797 (details cited by Richards, 2002, pp. 238–9).
16 Bumblebee Conservation Trust, 2013, p. 1.
17 Humboldt had read Thomas Malthus's 1798 *Essay on Population* which stated that population growth would outstrip food supply and that the only course of action to counteract this was sexual abstinence or birth control (McCrory, 2010, p. 27).

bordered by the Alps, and Lower Peru between the Pacific and the Andes, confirm this assertion. (Humboldt, 1995, pp. 150–1)

This unifying interconnectedness of Nature in which no species is isolated (including Man) is what we would now call ecology.[18] This is what Humboldt meant by the term 'cosmos'. In *Cosmos*, he reminds his readers that no species, whether plant or animal, are isolated. All things make up the web of life and even the mosquitoes that plagued him are 'an important point in the economy of nature':

> [We need] to recognize in the plant or the animal not merely an isolated species, but a form linked in the chain of being to other forms either living or extinct. They aid us in comprehending the relations that exist between the most recent discoveries and those which have prepared the way for them. Although fixed to one point of space, we eagerly grasp at a knowledge of that which has been observed in different and far-distant regions. (Humboldt 1997, p. 42)

Again in the above Humboldt passage a hint of Darwin can be seen in making a link 'between the most recent discoveries and those which have prepared the way for them', that is, changes that have taken place between the species in the past and the present. These changes show that the interconnectedness of things in Nature is more than a passive web of relationships that do not have an effect on each other; it is an active web made up of processes, interactions and transformations, and these are the same things that Darwin wanted to interrogate in his *Origin*. Humboldt's motto for this in *Cosmos* is 'everything is interaction' (*Alles ist Wechselwirkung*) (cited by McCrory, 2010, p. 172). Importantly Man needs to not only understand the connectedness of things in Nature between each other but their connections to the 'laws of the physical world':

> Man cannot act upon nature, or appropriate her forces to his own use, without comprehending their full extent, and having an intimate acquaintance with the laws of the physical world. (Humboldt, 1997, p. 53, also cited by Sachs, 2007, p. 78)

---

18   That is, the relationship between organisms and their environment.

Humboldt's belief in the unifying force of the interconnectedness of everything in Nature can be seen in his enthusiastic and scientific rigour of collecting samples of everything he was studying at the time to get a 'total picture' of the world.[19] This conviction of the interconnectedness of Nature was also reflected in his belief that all the academic disciplines of human knowledge should be connected. When delivering his inaugural Berlin Academy of Sciences speech in 1805 he predicted the 'interaction of all aspects of human knowledge harmonising in one organic whole' (cited by McCrory, 2010, p. 116). He believed that the study of one discipline would lead to another due to their interconnectedness. In addition to seeing this in Nature and in the disciplines, he also saw this interconnectedness between the scientists and strongly believed that they should work together as a team and share their knowledge[20] (McCrory, 2010, p. 116). This teamwork through 'organic' collaboration and the sharing of knowledge from different scientific disciplines is another example[21] of Humboldt's philosophy of the interconnectedness[22] of Nature, its disciplines, its naturalists and scientists. This can be seen in the writing of *Cosmos* in which he interrelates the work of other scientists such as John Herschel (McCrory, 2010, pp. 174–7). Darwin shared his knowledge in the same way with Charles

---

19 This would include notes, ideas, letters, newspaper cuttings, notes from colleagues, excerpts from books and lectures and would all be filed in boxes according to topics (McCrory, 2010, p. 106).
20 In 1828 Humboldt arranged the very first international Natural History convention. Goethe was invited, but could not attend (McCrory, 2010, p. 135).
21 That is, international 'scientific conferences' and international scientific collaborations are indebted to Humboldt for initiating this cooperation.
22 This view of the interconnectedness of nature was partly inspired by the Roman statesman and scholar Plinius (Pliny the Elder) who wrote *Natural History* in AD 77. Humboldt used a sentence from the text as a motto for his *Cosmos* on the title page: '*Naturae vero rerum vis atque majestas in omnibus momentis fides caret, si quis modo partes ejus ac non totam complectatur. – Plin., Hist, Nat., lib. Vii, c. 1.*' This translates as: 'The power and majesty of things in Nature lose their credibility if one's mind embraces parts of it only and not as a whole' (cited and translated by McCrory, 2010, p. 161).

Lyell,[23] Joseph Hooker[24] and many other scientists both at home and overseas.[25] Pre-eighteenth-century voyages were mainly concerned with trading in spices, silks and other goods. Humboldt (and Darwin after him), however, was now part of a group of European explorers such as Cook, La Condamine, Bougainville, Baudin and Malaspina who wanted the results of their scientific expeditions made available for the benefit of all.[26] The aim of this new group of explorers was to interrogate Nature to reveal her secrets in order to help society progress (McCrory, 2010, pp. 67–9; p. 130).

Humboldt's interrelatedness of Nature can be seen as developing Darwin's portrayal of Nature as organic, for example in his use of the 'web of affinities' in the *Origin*. This view of Nature is one that is self-purposive according to its needs without the need for a creative deity. He does this by merging the identity of God with Nature. The concept of an independent God creating Nature could now be seen as an unnecessary explanation as the natural processes of Nature are varied and develop themselves over long periods of time (Richards, 2002, p. 516 and p. 539). In Darwin's discussion of the 'conditions leading to dominant species and incipient species', he refers to organic constructs using such terms as 'diverse', 'competition', 'struggle', 'incipient species or varieties' and 'relationships' (Darwin, 1985, pp. 108–9). Darwin's presentation of causes and effects seen in Nature can be interpreted as coming from within Nature rather than being caused by a divine Creator.

---

23   In a letter to Lyell dated 19 February 1840, Darwin discusses his work on coral formation and the classification of reefs (Burkhardt, 2009b, p. 253).
24   In a letter from Hooker to Darwin dated 28 November 1843, Hooker refers to their discussion of Arctic and Antarctic flora and agreed for the plants, as well as those from the Galapagos, to be forwarded by Professor Henslow (who took receipt of them from Darwin when on the *Beagle*) (Burkhardt, 2009b, pp. 410–12).
25   Just one volume of Darwin's letters between 1837 and 1843 reveals approximately 600 references to those he either corresponded with or referred to in his correspondence. See the *Biographical register* in Burkhardt, 2009b, pp. 504–44).
26   Humboldt maintained contact with fellow scholars writing about two thousand letters a year. He shared his findings and encouraged others to make use of them, and this included Lyell, Hooker and Darwin who read his works (McCrory, 2010, pp. 131–3).

However, it is important to understand that Darwin did not always hold the view that Nature was self-purposive. Prior to going on the *Beagle*, he went along with the view that Nature had been designed by a Deity. Darwin had studied William Paley's *Natural Theology or Evidences of the Existence and Attributes of the Deity* (1802)[27] (Paley, 2005) in which God is depicted as a grand designer with a purpose, likened by Paley to a watchmaker:

> When we come to inspect the watch, we perceive – what we could not discover in the stone – that its several parts are framed and put together for a purpose, e.g. that they are so formed and adjusted as to produce motion. (Paley, 2005, p. 7)

For Paley the intricacies of the eye and the interrelated workings of the organs of the animal and human body showed that this was proof of intelligent design, just as the function and complexity of a watch imply a watchmaker. In his *Autobiography*, looking back at his time at Cambridge, Darwin states that he took Paley's arguments 'on trust' and did not question them, although he admits that this part of his BA course[28] 'was of the least use to me in the education of my mind' (Darwin, 1995, p. 18).[29] But when Darwin went on his voyage on the *Beagle* he began to question this view of Nature. His 'scientific' naturalist research was now looking for other explanations

---

27 In a letter to his cousin W. D. Fox dated 25–9 January 1829, Darwin refers to 'the little Go' (Burkhardt, 2009a, p. 74) which is an undergraduate name for the University 'Previous Examination', taken in the second year: 'The subjects of examination are one of the four Gospels or the Acts of the Apostles in the original Greek, Paley's Evidences of Christianity, one of the Greek and one of the Latin Classics' (*Cambridge University calendar*, 1829, p. 169). (Cited in Burkhardt, 2009a, footnote 7., p. 75).

28 It was fashionable for gentlemen of the time to study for an Arts degree covering such subjects as Natural Theology and Philosophy in order to become a clergyman and gain a living and then be able to follow one's interest in Nature by studying ornithology, collecting records of plant and animal species or writing Nature journals like Gilbert White before him.

29 'In order to pass the BA examination, it was also necessary to get up Paley's *Evidences of Christianity*, and his *Moral Philosophy*. This was done in a thorough manner, and I am convinced that I could have written out the whole of the *Evidences* with perfect correctness […]. *I did not at that time trouble myself about Paley's premises; and taking these on trust, I was charmed and convinced by the long line of argumentation*' (emphasis mine) (Darwin, 1995, p. 18).

that did not ask the question of how a Deity could be involved in the design of Nature. Rather than looking at what the prime cause of Nature was, Darwin was more interested in what the laws of Nature were and *how* they worked.[30] His theory of 'Natural Selection' proposed that animate and inanimate Nature 'evolved' and developed their laws without the *need* of a Creator creating them although he did not believe that it was impossible that a Creator could have created those laws responsible for the creation of Man. No wonder this contradiction made him feel 'bewildered'.[31] His *On the Origin of Species* (Darwin, 1985) is actually not an attempt to find the origin of all species (that is, the origin of life) but rather an examination of the process of transmutation or 'evolution' that leads to new species. In looking at the processes in this sense, Darwin was looking at the wider notion of what brought about diversity, which included races as well as species.

Darwin dealt with the problem of referring to 'some unknown plan of creation' by concentrating on the 'web of affinities', on the system that made up Nature rather than its cause:

> We can clearly see how it is that all living and extinct forms can be grouped together in one great system; and how the several members of each class are connected together by the most complex and radiating lines of affinities. We shall never, probably, disentangle the inextricable web of affinities between the members of any one class;[32] but when we have a distinct object in view, and do not look to some unknown plan of creation, we may hope to make sure but slow progress. (Darwin, 1985, p. 415)

---

30 In his introduction to the *Origin*, Darwin says that 'In considering the Origin of Species, it is quite conceivable that a naturalist [...] might come to the conclusion that each species had not been independently created, but had descended, like varieties, from other species. Nevertheless, such a conclusion, even if well founded, would be unsatisfactory, *until it could be shown how the innumerable species inhabiting this world have been modified* [...]' (emphasis mine) (Darwin, 1985, pp. 66–70).

31 In a letter to his friend Asa Gray written on 22 May 1860, Darwin says that although he does not believe in a God designing parasites or the eye, he says he can see no reason why Man should not have been created by laws created by a Creator and that he 'had no intention to write atheistically' (Burkhardt, 2009h, p. 224).

32 Present day DNA technology now makes it possible to 'disentangle the inextricable web of affinities' between many groups.

Showing that members within a class are related in some way makes any talk of a divine creator as a *necessary* cause redundant; this is not to say that relatedness within a taxonomic group precludes a divine creator, only that talk of a divine creator is not *necessary* when talking about Nature.[33] These tangled affinities make it difficult to isolate individual elements of Nature and therefore make it difficult to identify individual divine causes.[34] This in itself does not *necessarily* disprove a divine creator but makes it impossible to scientifically discover a first cause. What Darwin is saying here is that we only need to look at what is before us in Nature. What the observer can see is a 'web of affinities' making up a 'great system', and that is all that is required as a starting point to discovering the laws of Nature. For Darwin, therefore, there is no need to look to 'some unknown plan of creation'.[35]

The theory of Natural Selection in which diverse species and varieties transmute without the help of a Creator God, can be regarded as materialist. But as Darwin's 'tree diagram' shows (Darwin, 1985, pp. 160–1), everything in Nature is related to everything else through eventual shared common ancestry, and this concept of Nature can be seen as something more than just 'material' or 'physical'. Darwin described this interrelatedness as 'the inextricable web of affinities' (Darwin, 1985, p. 415) or the 'community of descent' (Darwin, 1985, p. 404).[36]

---

33 This point is made by Darwin when he argues against a deity creating species independently of each other: 'To admit this view is, as it seems to me, to reject a real for an unreal, or at least for an unknown, cause. It makes the works of God a mere mockery and deception' (Darwin, 1985, pp. 201–2).

34 That is, the notion of a divine power creating each species independently and the impossibility of scientifically establishing a separate divine cause responsible for each creation.

35 '[The] community of descent is the hidden bond which naturalists have been unconsciously seeking, and *not some unknown plan of creation*' (my emphasis) (Darwin, 1985, p. 404).

36 In this respect Darwin seems to be saying that the laws of Nature can be discovered not only by looking to the physical aspects of Nature but also from looking to something metaphysical in the relationships within the 'web of affinities'.

As Humboldt has already demonstrated, Nature is all about relationships within Nature; for example, how certain plants compete with each other for growing space, light and water; how they provide food for animals, and how both depend on bees to pollinate them to produce further seeds for propagation. The relationships in Nature are also about how changes in one species of plant or animal can lead to changes in another. Those species that are unable to adapt may become extinct. A reduction in rainfall, for example, can lead to greater competition between species that rely upon the same food source. The advantages of one species, through better adaptation and modification can therefore lead to the disadvantages and inevitable extinction in another, and likewise the disadvantages and extinction in one can lead to the diffusion of the other over a greater area and in greater numbers. These advantages are then inherited and passed on to the next generation, both in plants and animals. This perpetual flux and tension between relationships is aptly expressed through the terms 'flourishing', 'dominant', 'struggle' and 'modified':

> Hence it is the most flourishing, or, as they may be called, the dominant species, – those which range widely over the world, are the most diffused in their own country, and are the most numerous in individuals, – which oftenest produce well-marked varieties, or, as I consider them, incipient species. And this, perhaps, might have been anticipated; for, as varieties, in order to become in any degree permanent, necessarily have to struggle with the other inhabitants of the country, the species which are already dominant will be the most likely to yield offspring which, though in some slight degree modified, will still inherit those advantages that enabled their parents to become dominant over their compatriots. (Darwin, 1985, pp. 108–9)

This lush diversity of Nature is the one Humboldt had identified. For Darwin diversity depended on struggle in order for species to take advantage of their environment and their relationship to other species. The hardship of struggle forced change and modification. The term 'incipient species' is key here as it is a link in the chain between past and present, between an old form and a developing new form, and through this transformation-step represents all that is historical in Nature. Yet there is nothing permanent about it; it still has to 'struggle' to achieve its goal of becoming a fully-fledged species as it is competing against other incipient species doing the same thing. Those that succeed are those that have inherited the best advantages from their parents.

Darwin's organic concept of 'relationships' inspired by Humboldt is not only central to his argument of the freedom of Nature to create itself as product and creator, but is also used to demonstrate that, if species and varieties of species in a large genus are related to each other, then each species cannot be seen to be created by God. Darwin believes that if the characteristics of some species revert back to the characters of early progenitors, for example, a white stripe on a horse coming from its zebra forebears,[37] this is a reflection of a genealogical lineage developing over time and not one of God creating fixed species in time. The reversions are like faint watermarks running through a species revealing past and present genealogical structures, relationships and connections, with an inner core that remains structurally constant and an outer framework that is forever changing and developing. According to Darwin, development and diversity do not fit with the notion of 'a special act of creation' for if species were created by a God the assumption is that they would have been created for a special purpose.[38] If other species developed over time, this would show that the species created by God were not fulfilling their purpose, which would indicate that their original creation by God would have been created for 'no apparent reason':

> If we look at each species as a special act of creation, there is no apparent reason why more varieties should occur in a group having many species, than in one having few. (Darwin, 1985, p. 111)

Darwin's concept of 'relationships' is also emphasized in his principle of Divergence of Character which is one of the pillars that underpins his theory of natural selection. Species and varieties that develop those characteristics beneficial to their reproduction and survival will spread and reproduce the most and thereby pass these advantages down to their offspring.

---

37 Darwin, 1985, p. 201.
38 However, Richard Owen's view was that God created forms which *could* then change and develop and called this on-going event 'ordained continuous becoming' or, as Desmond and Moore put it, 'a sort of providential evolution'. In 1857 Owen presented the discovery of the hippocampus minor in humans as evidence to show that Man was different from a chimpanzee and so should belong to a special sub-class of mammal (Desmond and Moore, 2009, pp. 452–3).

This whole process is organic in that the tensions that make this possible come from within Nature and not from some unknown external cause such as God. In this sense selection is natural in that it does not come from God but comes from Nature, from its organic 'web of affinities'. Put simply, the closer varieties or species are related, or converge, the more limited their range[39] of spread; the more they differ or diverge, the greater their range of spread. This emphasizes the importance of organic relationships within Nature and demonstrates that convergence and divergence are the effects of changes within Nature due to its dynamic relationships, each acting on each other (as posited by Kielmeyer, mentioned earlier). It is the relationships that create Nature, not an unknown divine creator:

> Species very closely allied to other species apparently have restricted ranges. In all these several respects the species of large genera present a strong analogy with varieties. And we can clearly understand these analogies, if species have once existed as varieties, and have thus originated: whereas, these analogies are utterly inexplicable if each species has been independently created. (Darwin, 1985, p. 113)

The concept of change or development is central to Darwin's organic theory of natural selection. As already mentioned, reversion to characteristics of early progenitors reveals the hidden threads or vestiges of change 'over thousands on thousands of generations' (which Darwin alludes to when he talks about the incipient species which are varieties forming a bridge between species in the past and new species in the future). This tendency to produce the long-lost character Darwin calls the 'tendency hypothesis' and uses it to support his view that species were developed over time from primordial progenitors within Nature and not externally from God. Unlike Paley's intricate mechanisms put together by the divine watchmaker, all the relational parts are self-created by the parts that make up Nature. Through his powers of perception and aesthetic understanding through personal experience, Darwin is able to piece together these underlying relationships which only create the whole picture when put together. It is also Darwin's

---

39 In his *Origin of Species* Glossary, Darwin defines 'range' as 'The extent of country over which a plant or animal is naturally spread' (Darwin, 1985, p. 474).

use of his imaginative powers that enables him to penetrate the veil of Nature to discover its secrets; he frequently does this by arguing from analogy, that is, by making logical hypotheses based on personal experience or the experiences related by others in order to talk about the possible:

> I have stated that the most probable hypothesis to account for the reappearance of very ancient characters, is – that there is a *tendency* in the young of each successive generation to produce the long-lost character, and that this tendency, from unknown causes, sometimes prevails [...]. In several species of the horse-genus the stripes are either plainer or appear more commonly in the young than in the old [...]. For myself, I venture confidently to look back thousands on thousands of generations, and I see an animal striped like a zebra. (Darwin, 1985, p. 201)

Darwin then asserts that an argument stating that God created species with a tendency to vary does not explain anything; it does not explain the laws that make this happen or how they happen and is therefore not an argument at all (Darwin, 1985, pp. 201–2).

Darwin's self-purposive Nature, in which everything is related to everything else, can therefore be seen as a perpetual cycle of organic change feeding off itself creating difference through struggle, tension, adaptation and modification. The differences and diversification are created through natural selection. The relationship of species from different areas cannot be explained by the theory of independent creation but only through common progenitors and migration. Darwin refers to this when comparing the similarity of nearby island inhabitants to mainland Africa and South America showing that this does not support the view of independent creation (Darwin, 1985, pp. 385–6). Darwin's theory of natural selection promotes the view that species could not have been created uniquely and in isolation from other species, and, that through their development over generations and generations, they all share a common ancestry. The relationship between species from different areas and his examples of migration show that the independent creation of species cannot be possible.

Darwin's theory of natural selection encapsulates all that is organic and involves interdependent relationships, whether this be the increase and spread of one species and its varieties (its incipient species or intermediate forms), or the reduction and ultimate extinction of one in competition with its stronger more advantageous form. Therefore there can be 'no sort of explanation on the ordinary view of independent creation' (Darwin, 1985, p. 386). A more justifiable explanation is to look at the similarities of species in different areas as being due to colonization and then adaptation to their environment. Individuals having the best advantage and the most favourable variations then survive, and those having the most injurious get destroyed:

> Let it be borne in mind how infinitely complex and close-fitting are the mutual relations of all organic beings to each other and to their physical conditions of life [...]. Individuals having any advantage, however slight, over others, would have the best chance of surviving and of procreating their kind [...]. Any variation in the least degree injurious would be rigidly destroyed. This preservation of favourable variations and the rejection of injurious variations, I shall call Natural Selection. (Darwin, 1985, pp. 130–1)

Again, Nature is all about interdependent relationships between species, for example the bee relying on clover to make honey, as well as species competing with each other for resources, and species struggling for survival in their physical environment; all these demands and struggles necessitate modifications in order to secure an advantage over others, applying equally to young or adult animals and insects (although modifications in larvae may not be advantageous in the adult and therefore may not be carried over). Natural selection in the structure of the young (through the laws of correlation) will modify the structure of the parent and modifications in the parent will modify structures in the young but ensure they are not injurious for this would result in its extinction (Darwin, 1985, p. 135). In addition, species either prey or are preyed upon and are related to other species genealogically in time, all of which can be represented as a grand family tree. The dispersal and range of species depends on the range of other species. The look and feel of a geographical area (as experienced by Humboldt) therefore depends on these interdependent relationships: 'each organic being is either directly or indirectly related in the most important manner to other organic beings' (Darwin, 1985, p. 208).

The character of a geographical area therefore not only depends on the species, the terrain and the climate but the relationship between the species themselves: for example, the act of preying or being preyed upon, the competition between them, their numbers and range of habitat. These relationships are also reflected historically in time through fossil remains of extinct and incipient species, through reversions of characteristics in existing species and similarities of species between different regions showing migration. The tension and struggle in Nature between the strong and the weak is also reflected by distinct species that occupy large areas and intermediate varieties that occupy smaller areas and are in the process of becoming extinct:

> The two [distinct species] which exist in larger numbers from inhabiting larger areas, will have a great advantage over the intermediate variety, which exists in smaller numbers in a narrow and intermediate zone. (Darwin, 1985, pp. 209–10)

Under certain conditions, natural selection will modify the structure of beings to improve their advantage over others, but they do not necessarily produce the best structures under all conditions. So although the beings within a species may have improved themselves, thereby giving them an advantage over other species in the same habitat, they do not necessarily achieve absolute perfection. In a sense, there is a teleological pull towards improvement in that all successful species move along a developmental line, but the archetypal forms they seem to be moving towards cannot be known in advance as the struggles they face in the future are unknown. The unknown reflects the freedom of Nature to develop, produce and create in its own way that is most advantageous to itself. If everything were a simple act of creation by the divine Creator, everything would be fixed. Yet Darwin notes animals that have changed their habits without changing their structures; for example, woodpeckers that live where there are no trees or upland geese with webbed feet that rarely go near water. Rather than bring in a divine Creator as the cause of these anomalies in Nature, it is easier to explain them as a result of a struggle where creatures have had to move on to new habitats but have not found it injurious to keep their present structures (the effort of change and the corresponding use of resources is only used if it is absolutely necessary for its survival and advantage over other species):

> If about a dozen genera of birds had become extinct or were unknown, who would have ventured to have surmised that birds might have existed which used their wings solely as flappers, like the logger-headed duck (Micropterus of Eyton); as fins in the water and front legs on land, like the penguin; as sails, like the ostrich; and functionally for no purpose, like the Apteryx. Yet the structure of each of these birds is good for it, under the conditions of life to which it is exposed, for each has to live by a struggle; but it is not necessarily the best possible under all possible conditions. (Darwin, 1985, p. 214)

As already demonstrated, Darwin had been influenced by Humboldt's organic view of Nature in which the conditions leading to species-change came from inter-relational causes and effects *within* Nature, a Nature that is both producer and product. Change occurs within Nature due to species, habitat, climate, food sources, etc. all acting on each other. Where species have improved their structure for living in a certain habitat their intermediate stages will have been replaced by the improved form, and so the intermediate forms will have become extinct. As there will have been less of them there will be less chance of their being discovered as fossils. Darwin therefore has no problem with the possibility of an insect-eating bear taking to water developing into a whale (misunderstood by his readers as happening over a short period of time and therefore being a ridiculous idea):

> In North America the black bear was seen by Hearne swimming for hours with widely open mouth, thus catching, like a whale, insects in the water. Even in so extreme a case as this, if the supply of insects were constant, and if better adapted competitors did not already exist in the country, I can see no difficulty in a race of bears being rendered, by natural selection, more and more aquatic in their structure and habits, with larger and larger mouths, till a creature was produced as monstrous as a whale. (Darwin, 1985, p. 215)

The diversity in the number of species conforms to Darwin's theory of natural selection in which one species, through modification, gives rise to two or more varieties (or 'incipient species'), which are then converted into species in turn producing further species[40] 'like the branching of a great tree from a single stem' (Darwin, 1985, p. 321).

---

40 This is not to suggest that species have been forever creating further species with the number of species getting larger and larger. Natural selection means that those species

This 'web of affinities' includes life and death, whether this is 'budding' life reproducing itself, or 'dead and broken branches' representing extinct species:

> The affinities of all the beings of the same class have sometimes been represented by a great tree [...]. The green and budding twigs may represent existing species; and those produced during each former year may represent the long succession of extinct species [...]. Of the many twigs which flourished when the tree was a mere bush, only two or three, now grown into great branches, yet survive and bear all the other branches; so with the species which lived during long-past geological periods, very few now have living and modified descendants [...]. As buds give rise by growth to fresh buds, and these, if vigorous, branch out and overtop on all sides many a feebler branch, so by generation I believe it has been with the great Tree of Life, which fills with its dead and broken branches the crust of the earth, and covers the surface with its ever branching and beautiful ramifications. (Darwin, 1985, pp. 171–2)

The image of the tree is a useful one as it can incorporate the living branches with the dead branches. The dead branches show where the living ones have come from and the living ones show the present results of past transformations. This is life and death in equal balance, as equal partners in the creation process. This mix of life and death can be seen in the words used from the passage above, such as 'green and budding', 'extinct', 'existing', 'produced', 'flourished', 'grown', 'survive', 'bear', 'lived', 'vigorous', 'branch out and overtop', 'feebler' and 'dead and broken'. This argument for an organic Nature forever recreating itself is repeated again and again throughout the *Origin*. All the parts of Nature through their relationships and influences create life and death through modifications to maintain survival and continued reproduction. Central to this is 'connection' between the parts making up the 'grand system':

> We can understand how it is that all the forms of life, ancient and recent, make together one grand system; for all are connected by generation. (Darwin, 1985, p. 342)

---

that are the most successful through their modifications will survive and multiply, but those that are not successful will go into decline and eventually become extinct. So with increase, there is also decrease.

Darwin also uses the image of an 'entangled bank' to express the tension and mutual relationships between the life forces in Nature. On one of his walks Darwin noticed how a bank was filled with competing life, consisting of bushes, plants, snails and birds. This concept of the entangled bank and its entangled relationships is spelt out in the very last paragraph of the *Origin*:[41]

> It is interesting to contemplate an entangled bank, clothed with many plants of many kinds, with birds singing on the bushes, with various insects flitting about, and with worms crawling through the damp earth, and to reflect that these elaborately constructed forms, so different from each other, and dependent on each other in so complex a manner, have all been produced by laws acting around us. These laws, taken in the largest sense, being Growth with Reproduction; Inheritance which is almost implied by reproduction; Variability from the indirect and direct action of the external conditions of life, and from use to disuse; a Ratio of Increase so high as to lead to a Struggle for Life, and as a consequence to Natural Selection, entailing Divergence of Character and the Extinction of less-improved forms. Thus, from the war of nature, from famine and death, the most exalted object which we are capable of conceiving, namely, the production of the higher animals, directly follows. There is grandeur in this view of life, with its several powers, having been originally breathed into a few forms or into one; and that, whilst this planet has gone cycling on according to the fixed law of gravity, from so simple a beginning endless forms most beautiful and most wonderful have been, and are being evolved. (Darwin, 1985, pp. 459–60)

This last paragraph articulates the concept of organic Nature on two levels. First on a simple level, it describes a bank in nature overgrown with bushes providing a habitat for wildlife all competing with one another in various ways yet furthering life: worms breaking up the earth helping plants grow but at the same time providing a source of food for the birds. All species in Nature are in relation to their habitats and each other form a unified whole. Through their adaptations to each other and their environment they shape the colour and texture of their existence, whether they are competing against each other or working together to create a mutually beneficial state

---

41   This last paragraph of Darwin's *Origin* is 150 years before its time in summing up modern ecology as well as the central importance of evolution in the understanding of life and death in Nature. With these points in mind there is indeed a 'grandeur in this view of life'.

of Nature. This is Humboldt's unified Nature. But at another more complex level, the concept of organic Nature is also the interrelationship of the laws of Nature that make this possible: namely, 'Growth', 'Reproduction', 'Inheritance', 'Variability', 'Ratio of Increase' leading to a 'Struggle for Life',[42] and, consequently, 'Natural Selection entailing Divergence of Character and Extinction of less improved-forms' (Darwin, 1985, pp. 459–60). These constructs that make up Darwin's theory of natural selection form an interrelated chain, each link being a part of the whole, collectively and necessarily creating the power or force of Nature through its core forms or archetypes. There is difference and diversity between the species and the individuals that make up the species, and there is difference and tension between the laws, yet they are all 'dependent on each other in so complex a manner' (Darwin, 1985, pp. 459–60). There is an intricate, finely balanced tension between divergence and extinction, between life and death; an increase in life on one side, creating a decrease on the other.

Viewing Nature as an organic whole made up of its parts creates the view of a single reality. There are no hidden mechanisms creating Nature or its laws *outside* Nature. Any laws or mechanisms are parts of Nature, and, as such, contribute to making up the whole. Humboldt's oneness of Nature can be seen in *Cosmos*.[43] In this work he captures every aspect of natural science in the whole universe. Humboldt used the term 'Cosmos'

---

42 When discussing Darwin's language of 'struggle' and 'competition', it is worth considering Karl Marx's view, in addition to Humboldt's view of organicism, that what Darwin discovers in Nature and informs his theory of natural selection is a mirror image of what exists in society, namely, capitalism. In 1862, Karl Marx, in a letter to his collaborator Friedrich Engels, wrote: 'It is remarkable how Darwin recognises among beasts and plants his English society with its division of labour, competition, opening up of new markets, "inventions", and the Malthusian "struggle for existence". It is Hobbes' "bellum omnium contra omnes" ["the war of all against all"]' (Radick, 2003, p. 143). However, Marx and Engels saw Malthus as 'a bourgeois fraud' as they believed that it was the political system rather than natural laws that created poverty and prevented progress (Young, 1985, p. 53). Marx sent Darwin a copy of his *Das Kapital* in 1873 with the inscription 'Mr. Charles Darwin on the part of his sincere admirer Karl Marx', but it remained uncut and unread (Browne, 2003, p. 403).
43 Published in five volumes between 1845 and 1862.

to combine the physical laws of the universe with the terrestrial world of biology and chemistry and in this sense 'Humboldt's cosmology brought spirituality down to earth' (Sachs, 2007, p. 75).[44] Humboldt wanted his readers to recognize that despite the difficulty of discovering the laws of biology compared with the mechanical sciences, there was a certain order in the terrestrial and celestial realms and that just in the same way as the earth and sky were one, so too were human beings a part of this union. In his Berlin lectures, Humboldt wanted his audience to be freely 'astonished' by the Cosmos, thereby creating a society free from authoritarianism and Christian orthodoxy. It is important to mention Emerson in this context as there is evidence to show that Humboldt got the concept 'Kosmos' from Emerson. In his book *Nature*, in 1836, Emerson developed his form of transcendentalism in which Man can regain his loss of spirituality through the contemplation of Nature in solitude (Sachs, pp. 73–4). Like Emerson, both Humboldt and Darwin were aware how Nature leaves evidence of its work in rock and species, and that Man as he makes his own journey of life through Nature is able to view his own past connections and leave a further trail both individually and as a species (a form of 'rastro'). For both Man and Nature this is the history of Nature or 'Natural History'. Emerson[45] aptly describes this when he says:

> We walk on molten lava on which the claw of a fly or the fall of a hair makes its impression, which being received, the mass hardens to flint and retains every impression for evermore [...]. Is it not better to intimate our astonishment as we pass through this world if it be for a moment ere we are swallowed up in the yeast of the abyss? I will lift up my hands and say 'Kosmos'. (Ferguson, 1964, 4:16; cited by Sachs, 2007, p. 74)

This oneness can be seen as both the impression and the thing making it, both contributing to its history. In Darwin this unity can be seen in both

---

44 Humboldt's use of the word 'Cosmos' came from Emerson who took it from the Greek meaning 'celestial bodies'. Emerson lost his religious faith after the death of his wife, Ellen, in 1830 but regained his spirituality through Nature (he read Humboldt's *Personal Narrative*) (Sachs, 2007, p. 75).

45 When Emerson lifted up the tomb of his wife and saw her ashes, he realized the importance of contemplating Nature as we make our way through life (that is, by being 'astonished' by it) (Sachs, 2007, p. 74).

the physical instance of an individual tree and its history through the line of its species, but also in the metaphorical tree of life (another 'rastro') representing the development of all species in space and time, both past and present. For Humboldt this oneness or 'Cosmos' is about 'mutual dependence and connection' and about the mysterious occult relations that exist between its parts. The haunting melody of Humboldt's favourite *capirote* bird on the Canary Islands would not be possible without a particular plant the bird lives off which only grows on volcanic soil at a particular altitude (Sachs, 2007, p. 76).

Humboldt sums up this communion with Nature as 'order and harmony' and the contrast 'between the narrow limits of our own existence' and infinite Nature. This goes to the heart of Darwin's theory of natural selection – these laws are natural laws developed by Nature and in harmony with Nature. They include the development and the demise of species through selection, thus representing both infinity and 'the narrow limits of our own existence' both individually and as a species. This unity expressed by Humboldt is a hint of a theme later taken up by Darwin:

> The earnest and solemn thoughts awakened by a communion with nature intuitively arise from a presentiment of the order and harmony pervading the whole universe, and from the contrast we draw between the narrow limits of our own existence and the image of infinity revealed on every side, whether we look upward to the starry vault of heaven, scan the far stretching plain before us, or seek to trace the dim horizon across the vast expanse of ocean. (Humboldt, 1997, p. 25)

Darwin finds no evidence of a Creationist explanation (of Divine creation) in Nature. The only explanation is that of natural selection which is the product of Nature's own creation. For Darwin, rudimentary, atrophied or aborted organs do not offer an explanation for the Creationist view of Nature as this does not give an explanation but only restates fact (Darwin, 1985, p. 430). The evidence of Nature recreating itself can be seen in living species as well as their fossilized remains; as in a stick of rock, the inner writing runs through time and is the history of time, the history of Nature. The rudimentary, atrophied or aborted organs are the traces of this history (these traces are another 'rastro').

In support of this, Darwin gives examples of snakes with rudimentary pelvis and hind limbs, beetles with mere rudimentary membranes as wings, and teeth in the upper jaws of unborn calves. When reading these examples, it must be remembered that the argument for there not being a Creator having a hand in the creation or development of species is not a theological debate about the existence of God. It is an argument stating that the laws of Nature exhibited through natural selection are forces within Nature; that Nature is both creator and product. It is an argument presenting a unity of Nature which is organic; a unity in which all the parts are interrelated and make up the whole; a Nature which is one reality. A number of Romantic[46] British poets before Darwin have taken up this theme, such as Samuel Taylor Coleridge and William Wordsworth. For example, in Coleridge's *The Eolian Harp* (1795), the lines, 'And what if all of animated nature/Be but organic Harps diversely framed,/That tremble into thought, as o'er them sweeps/ Plastic and vast, one intellectual breeze,/At once the Soul of each, and God of all?' (Coleridge, 2013, lines 45–9) reflect the idea of Nature being composed of music created by a collection of 'organic Harps', 'one intellectual breeze', one Soul or God (and like Humboldt and Darwin), gaining an understanding of the essence of Nature through the imagination (although unlike Humboldt, Coleridge's vision is implicitly Unitarian, and hence inspired by religious thought). In Wordsworth's 'Love of Nature Leading to Love of Mankind' (1805),[47] the poem shows how, in different ways, the external Nature experienced by Man and his own nature are one reality as this is experienced through his senses (the object experienced and the sense of experiencing it through the senses makes it one – an unknown Nature does not exist independently of the observer's senses as it can do in Kant).[48] This oneness with Nature

---

46 This thematic link to Wordsworth's concept of 'one reality' could be used to support the claim that both Humboldt's and Darwin's notion of 'one reality Nature' is Romantic.
47 This poem is part of *The Prelude* (Wordsworth, 1970, Book viii). It was unpublished during Wordsworth's life-time, and was not read by Darwin during his formative youth.
48 This independence of the thing-in-itself, independent of 'necessity' and 'appearances', is expressed by Kant in his *Critique of Pure Reason* (1769–1780) as follows: 'Inasmuch as it is *noumenon*, nothing *happens* in it; there can be no change requiring dynamical determination in time, and therefore no causal dependence upon appearances. And

could also be seen as similar to Humboldt's aesthetic experience of Nature through the senses, and similar to Darwin's experience of Nature through his 'imagination'. One section of Wordsworth's poem reminds the reader of Plato's 'The Simile of the Cave' in *The Republic* (Plato, 1971, pp. 278–86) in which the prisoners (imprisoned since childhood) cannot distinguish the shadows from the objects in reality. As Wordsworth puts it:

> Substance and shadow, light and darkness, all
> Commingled, making up a Canopy
> Of Shapes and Forms and Tendencies to Shape
> That shift and vanish, change and interchange
> Like Spectres, ferment quiet and sublime.
> (Wordsworth, 1970, Book viii, lines 719–23)

Just as a traveller learns about Nature and life through his experiences (and needs to distinguish between the real and the unreal, the substance and the shadows), Man (as a traveller through life) interprets Nature through his senses, and as he matures in time he learns (through knowledge gained through 'past and present') to understand the individual characteristics of Nature and the interconnected whole. Like Darwin and Humboldt who discover the power of Nature creating itself, so too with Wordsworth who (looking back at his youth) finds the power of Nature in things and in Mankind:

> I sought not then
> Knowledge; but craved for power, and power I found
> In all things; nothing had a circumscribed
> And narrow influence; but all objects, being
> Themselves capacious, also found in me
> Capaciousness and amplitude of mind.
> (Wordsworth, 1970, Book viii, lines 754–9)

---

consequently, since natural necessity is to be met with only in the sensible world, this active being must in its actions be independent of, and free from all such necessity. No action begins *in* this active being itself; but we may yet quite correctly say that the active being *of itself* begins its effects in the sensible world' (Smith, 1970, p. 469).

Although on the one hand Wordsworth can be seen to regard shepherds and country folk (common man) as the closest to Nature and therefore nobler than those of the cities having developed fewer vices (as with Rousseau's 'noble savage'), he is nevertheless inspired by the experience of London which can both develop knowledge in the individual and create a unity of humanity. Just as Darwin saw a unified Nature in which all races of Man came from the same origins and should therefore be treated with the same humanity, so too did Wordsworth see 'the unity of man' (through his inspiration from the people of London) with a single 'spirit', a single 'moral sense', a 'soul' that is a part of God breathing through the whole of Nature:

> Of that great City, oftentimes was seen
> Affectingly set forth, more than elsewhere
> Is possible, the unity of man,
> One spirit over ignorance and vice
> Predominant, in good and evil hearts
> One sense for moral judgements, as one eye
> For the sun's light. When strongly breath'd upon
> By this sensation, whencesoe'er it comes
> Of union or communion doth the soul
> Rejoice as in her highest joy: for there,
> There chiefly, hath she feeling whence she is,
> And, passing through all nature rests with God.
> (Wordsworth, 1970, Book viii, lines 825–36)

However, unlike Wordsworth's form of God breathing life through Nature or 'Nature [resting] with God', Darwin questions God's single-handed creation of Nature by showing that species-change is a slow and gradual process and that therefore it is impossible for a God to have created all species at the same time. This is demonstrated by fossil remains being more closely related the closer their formations are to each other; the more distant they are from each other the more they will differ, showing that they have changed over time (Darwin, 1985, p. 440).

Darwin's view that the production and extinction of inhabitants, both past and present, is due to 'secondary causes' rather than a Divine first cause accords with the empirical view that Nature is both product and creator.[49] Fossils indicate the chain of descent running through Nature rather than pointing to a realm beyond Nature. The existence of fossils of intermediate groups found between formations demonstrates their intermediate positions in the chain of descent indicating their descent and the descent of extinct species from common parents (Darwin, 1985, p. 448). Also the existence of allied species from two different areas shows that they must have come from the same parents. This does not support the theory of creation (Darwin, 1985, p. 450) and again reflects the notion of Nature through the wholeness of its parts (Nature creating itself), its interconnected diversity and mutual interdependence of struggle for existence and survival. In a sense, this going back to the same beginnings is a paradox since it is expressed as the opposite of diversity; yet going forward through space and time in the face of struggle creates the development of diversity and adaptation (as well as extinction). Aesthetically[50] this diversity is represented in the sketches made by naturalists and reproduced in their plates;[51] the images represent an intimate union and relationship between the observed and the observer, between the intellectual ideas of the naturalist and the species and individuals that make up the tapestry of Nature. The sketches and experiences of Nature capture its wholeness and in so doing

---

49 Darwin does not actually personify the laws of natural selection as a 'creator' but his argument is that if Nature is both creating species through its laws and at the same time is the product of its laws, it could be seen as the personification of a God-like creator (but yet not a God).

50 Aesthetically rather than visually since the sketches and plates represent the interrelationship of the viewer and the object. The representations are not merely factual but also represent how the subject perceives Nature using the 'Humboldtian Method'.

51 The first permanent photograph was not made until 1827 by Niépce (Bellis, 2013, p. 1). Naturalists in the early nineteenth century, therefore, had to send specimens back to their respective countries as proof of what they had observed. However, the colour of pressed plants would fade and specimens would get damaged. Taking sketches and writing notes of their observations was at the time the only way of making a permanent record, and this information was important for the lithographers making the plates for the naturalists' publications.

bring out the spirit of the relationship between naturalist and Nature (its inner secrets and its outer appearances).

This chapter has provided strong evidence to suggest that Humboldt was one of the most important naturalists of his time to have exerted a significant influence on Charles Darwin, whether this be in Darwin's thought processes, his texts or his research methods. Humboldt's concepts of unifying processes and interrelationships in Nature making it 'organic' can be seen in Darwin's examples of struggle and mutuality found in the 'entangled bank' of Nature from as far afield as the Galapagos to his own garden at Down House. Humboldt's concept of the 'one reality of nature' can be seen in Darwin's theory of natural selection in which Nature creates itself through its own natural laws. Species selection reflects Nature's split between the finite and the infinite, between diversity and extinction over time in which individuals and individual species die off due to more successful competitors that are better at adapting and diversifying. Although Darwin's antipathy towards slavery was developed at an early age through his free-thinking family and relations, he was additionally influenced by his reading of Humboldt who also believed in a common humanity (like Wordsworth). This belief that Man was part of one community can be seen in his view that all men, regardless of race, came from the same origins.

CHAPTER 2

The Forces of Nature in Humboldt and Darwin

This chapter looks at the importance that the direct experience of the forces of Nature played in developing both Humboldt and Darwin's aesthetic imagination. For Darwin, these insights into the workings of Nature, together with the inspirational insights from Humboldt, enabled him to use his imagination to explore the hypotheses that led to his theory of natural selection. Their insights also led to their belief that all of mankind are born equal and this chapter will also examine their abhorrence of slavery showing that Man's 'common humanity' can also be a force of Nature determining the perpetuation of Man's existence.

One of the fascinations that drew naturalists and chemists in the late eighteenth century towards the desire to discover the secret forces of Nature was the discovery of electricity. By the time Humboldt was twenty-six, in 1795, he was already very interested in the experiments in electricity conducted by Galvani and Volta. Galvani had demonstrated that you could make frogs legs convulse by applying two different metals to the muscles and nerves of the leg. Galvani believed this was because the electricity was contained in the nerves, calling it 'animal electricity', publishing his experiments in *Commentary on the Effect of Electricity on Muscular Motion* (1791). He believed 'animal electricity' was an additional form of electricity to natural electricity such as lightning and 'artificial electricity' such as friction (static electricity). Volta (who coined the term 'galvanism'), on the other hand, believed that the electricity was created by the contact between the two metals. Humboldt conducted electrical experiments on himself but came up with his own hypothesis that 'the metals intensified the convulsions but were not the cause of them' (Botting, 1973, p. 33 and McCrory, 2010, p. 41). Humboldt published the results of 4,000 experiments in his *Experiments on the excited muscle and nerve fibre with conjectures on the chemical process of life in the animal and vegetable world* (1797). But his hypothesis was discredited when Volta published his own experiments showing how electricity could be produced without any tissue by putting two metals together with a damp cloth or liquid (Botting, 1973, p. 33). The experiments Humboldt conducted on himself show how intensely he experienced the forces of Nature and the extreme measures he took in order to experience them regardless of pain, or risk to life or limb. One particular experiment is worth quoting at length to make this point:

> I raised two blisters on my back, each the size of a crown-piece and covering the trapezius and deltoid muscles respectively. Meanwhile I lay flat on my stomach. When the blisters were cut and contact was made with the zinc and silver electrodes, I experienced a sharp pain, which was so severe that the trapezius muscle swelled considerably, and the quivering was communicated upwards to the base of the skull and the spinous processes of the vertebra. Contact with silver produced three or four single throbbings which I could clearly separate. Frogs placed upon my back were observed to hop. Hitherto my right shoulder was the one principally affected. It gave me considerable pain, and the large amount of lymphatic serum produced by the irritation was red in colour and so acrid that it caused excoriation in places where it ran down the back. The phenomenon was so extraordinary that I repeated it. This time I applied the electrodes to the wound on my left shoulder, which was still filled with a colourless watery discharge, and violently excited the nerves. Four minutes sufficed to produce a similar amount of pain and inflammation with the same redness and excoriation of the parts. After it had been washed my back looked for many hours like that of a man who had been running the gauntlet.[1] (Cited by Botting, 1973, p. 34)

This experiment shows Humboldt's almost obsessive desire to experience the raw forces of Nature within his very own body so that he is almost indistinguishable from them. It is as if by experiencing the pain, the objective cause and the subjective affect become one (and as referred to in the introduction, this is similar to the Romantic tradition of the landscape artist both describing the scene objectively as the observer but also being present in the picture as a subject).The pain is not described in a personal way but in a way that is defined objectively, as if it were a part of the natural force observed by him through seeing the muscles swell. The number of throbs

---

[1] Unfortunately no references are given for this quote.

are counted rather than described as horrible. Although his shoulder was in pain he is more interested in the production of the serum. He even repeated it despite the pain. The use of the frogs hopping on his back underlines this almost mad obsession with wanting a direct, unfiltered communion with the secret forces of Nature. The experiments of Galvani and Volta undoubtedly inspired a fascination in the chemical processes of life and this in turn made him want to study plant and animal life and what their similarities and differences were (McCrory, 2010, p. 41). Humboldt was familiar with Priestley's 'phlogiston theory',[2] and, especially, Lavoisier's[3] work, having read his *Traité élémentaire de chimie* (1789) three times, and this added to his enthusiasm in the study of the hidden forces of Nature[4] (McCrory, 2010, p. 53). Darwin would have been familiar with these experiments not only through his reading of Humboldt, but also through the work and life of his grandfather Erasmus Darwin, who, through his membership of the Lunar Society, was an ardent follower of Galvanism (Uglow, 2003, pp. 428–9).

These measurements of the forces of Nature in Humboldt's own body are again referred to in his *Personal Narrative* when he ascends Mount Chimborazo, the highest ascent up a mountain anyone had made at that time. He had no special clothing or equipment as climbers would have today. He had to climb along narrow ledges with bottomless depths – one false move and he would fall thousands of feet. At the same time he experienced the tremors of the erupting volcano and took measurements with his instruments on the lip of the crater. On 23 June 1802 he climbed to a height of 20,702 feet above sea-level, experiencing nausea, giddiness, breathlessness, bleeding of the lips and gums and bloodshot eyes:

---

2   According to this theory, Phlogiston was a substance created during combustion.
3   Lavoisier repeated Priestley's experiments and discovered that the active 'principle' of the atmosphere was oxygen and was present in both the processes of combustion and respiration, that is, oxidization.
4   Humboldt gave demonstrations of Galvanism on the nerves of frogs whilst in South America. He was also amazed to see the electric eels that stunned and resulted in the drowning of horses used to fish them (Humboldt, 1995, pp. 170–1).

> But in the same individual they constitute a kind of gauge for the amount of rarefaction in the atmosphere and the absolute height he has reached. (cited by Botting, 1973, p. 153)[5]

As the above experiments on Humboldt's own body indicate, the natural forces and processes of Nature can create both pain and amazement. So too can the experience of jungles create the feeling of disorientation and beauty. The wonders of Nature with all its diversity, whether this be of species or of climate and terrain, create the image of a Garden of Eden, with both its beauties and life-threatening dangers; whether this be the amazing variety of plants, colours, aromas and wildlife, or the dangers of poisonous snakes, crocodile-infested rivers or clouds of mosquitos (McCrory, 2010, p. 74). The forces of Nature are not all for good and its beauty is a veneer covering a darker side:

> You find yourself in a new world, in a wild, untamed nature. Sometimes it is a jaguar, the beautiful American panther, on the banks; sometimes it is the hocco (Crax alector) with its black feathers and tufted head, slowly strolling along the sauso hedge. All kinds of animals appear, one after the other. 'Es como en el paraíso' ('It is like paradise') our old Indian pilot said. Everything here reminds you of that state of the ancient world revealed in venerable traditions about the innocence and happiness of all people; but when carefully observing the relationships between the animals you see how they avoid and fear each other. The golden age has ended. In this paradise of American jungles, as everywhere else, a long, sad experience has taught all living beings that gentleness is rarely linked to might. (Humboldt, 1995, pp. 178–9)

This mixture of innocence and viciousness in which Nature can be 'at once wild and tranquil, gloomy and attractive' is a reminder that although the world of the jungle is different from our own, it 'is where we come from' (Sachs, 2007, pp. 62–3). Humboldt not only experienced this darker side in the form of mosquitoes, dangerous seas where he nearly lost his life or even the smoking volcanoes that had already taken many lives, but in the way Man treated Man in the form of slaves:

---

5   Unfortunately no reference is given for this quote.

It is distressing to think that still today in the Spanish West Indies slaves are branded with hot irons to identify them in case they escape. This is how one treats those 'who save other men from the labour of sowing, working in the fields and harvesting'. (Humboldt, 1995, p. 67)

Both Humboldt and Darwin abhorred slavery and their voices echo Rousseau's when he says (in 1762) that 'Man was born free, and everywhere he is in chains' (Rousseau, 2008, p. 45) and 'To renounce our freedom is to renounce our character as men, the rights, and even the duties, of humanity' (Rousseau, 2008, p. 50). Humboldt learned from Rousseau[6] that travelling, especially in South America, was the best way to 'shake off the yoke of national prejudices' (Wilson, 1995, p. xlix). Darwin could understand the struggle between different species of animal but not between Man and Man, as his moral upbringing,[7] learning taxidermy from a 'blackamoor' whilst at Edinburgh University, travelling with native guides and seeing first-hand the appalling treatment of slaves in South America, had taught him to regard all men as *potentially equal*;[8] for Darwin, regardless of race, all men came from a common progenitor.[9] Later in life when writing the

---

6   From reading Rousseau's *A Discourse on Inequality* (1755) in which he says that the book of Nature 'never lies' (Wilson, 1995, p. xlix).
7   His whole family and his wife Emma's Wedgwood family were actively against the slave trade. The Wedgwoods gave financial support to the Anti-Slavery Society (Desmond and Moore, 2010, pp. 60–1).
8   Although Darwin can historically be regarded as a man of his time, by today's moral standards a certain degree of racism and sexism can be detected in his writings. For example, sexism can be seen in the *Descent of Man* when comparing the mental powers of Man and Woman: 'Woman seems to differ from man in mental disposition, chiefly in her greater tenderness and less selfishness; and this holds good even with savages [...]' (Darwin, 2004, pp. 629–31). Racism can be seen in *The Descent of Man* when Darwin expresses the view that some races such as the 'negro' or 'Australian' are nearer to the ape than 'man in a more civilized state' (Darwin, 2004, pp. 183–4; also Lyons, 2013, p. 1).
9   'All the races agree in so many unimportant details of structure and in so many mental peculiarities, that these can be accounted for only by inheritance from a common progenitor;' (Darwin, 2004, p. 678). That all men and women were *potentially equal* did not mean that Darwin regarded them as equal in his lifetime; only that they *could* rise up to develop *his* and *his society's* definition of what it was to be a cultivated

first volume of *Cosmos* (1845), Humboldt reiterates the assumption that all men are born equal but cultivate themselves according to where men find themselves:

> While we maintain the unity of the human species, we at the same time repel the depressing assumption of superior and inferior races of men. There are nations more susceptible of cultivation, more highly civilized, more ennobled by mental cultivation than others, but none in themselves nobler than others. All are in like degree designed for freedom. (Humboldt 1997, p. 358)

Humboldt acknowledges that not all men treat each other as equal and that therefore humanity is not unified but he believes that men should work together to try to achieve it. At the end of *Cosmos* Humboldt quotes directly from his brother Wilhelm the goal of achieving a 'common humanity':

> It is that of establishing our common humanity – of striving to remove the barriers which prejudice and limited views of every kind have erected among men, and to treat all mankind, without reference to religion, nation, or colour, as one fraternity, one great community [...]. This is the ultimate and highest aim of society, identical with the direction implanted by nature in the mind of man toward the indefinite extension of his existence. (Humboldt 1997, p. 358)

'The direction implanted by nature in the mind of man' is part of that force in Nature that determines the perpetuation of Man's existence. For this to work Man needs to listen to the forces of Nature that govern his existence to ensure that his actions mirror them. To do this he needs to understand his 'common humanity' and that Man is part of 'one great community'.[10] There are many paradoxes here concerning the forces or laws of Nature that

---

individual. As summed up by Desmond and Moore: 'For it was *his* class Darwin was talking to: 'savages' and the 'uneducated' knew what they liked but lacked any higher aesthetic refinement. Again, he was intimating the steps of evolutionary ascent through savagery towards a Victorian gentleperson's taste for the beautiful, whether in landscapes, music or adornment' (Desmond and Moore, 2010, p. 374).

10   Humboldt practised what he preached. On the ascent of Chimborazo, he refused to be carried by the *cargueros*, the Indian man-carriers or human mules. Despite having injuries to his feet, he preferred to suffer himself than to see others suffer (Botting, 1973, p. 149).

determine his nature. On the one hand, as Darwin shows through his 'tree' of life, all races of Man come from a common progenitor.[11] In this respect, despite adapting to their own environments, they are all equal in terms of their 'common humanity'. Yet, as in the animal world, there is that darker side of humanity in which Man fights against Man to take advantage, and slavery is an example of this. Although Man is mentally aware of his actions, his freedom to choose can be used to distort the facts to justify actions to his advantage. In the case of natives their 'primitive' cultures are demonstrated as proof that their cultures are 'inferior' to European civilizations, giving the Europeans the right to use them as slaves. Even in the 1820s in Jamaica and Brazil black people were regarded as a separate species 'merely an intermediate step between man and the brute creation' (Desmond and Moore, 2010, p. 10). This view was also reinforced by 'phrenology' in which the size of a person's head was said to determine their intelligence, meaning that education played little part in the development of a person's mental abilities.[12] Yet naturalists such as Humboldt and Darwin and political reformers such as William Wilberforce demonstrated that there was also a moral and enlightened dimension to Man's reasoning.[13] Just like Darwin's 'web of affinities', Humboldt believed in 'the ties of consanguinity', that is, the descent from the same ancestor. These ties link together those races speaking different languages with different cultures. Humboldt wanted his readers to look beyond those differences 'to discover some of those family

---

11   Darwin's breeding experiments with pigeons showed they came from a common stock and that the different pigeons were different varieties rather than different species. By analogy Darwin believed the same applied to human races as they could interbreed without causing sterility as is generally the case between different species. As summarized by Desmond and Moore (2010, p. 258): 'The pigeon family tree became a Darwinian paradigm and a perfect metaphor for mankind's family tree'.
12   George Combe was a leading exponent of phrenology.
13   Yet Darwin finds it quite acceptable for Indians to be butchered in warfare if it means a more progressive civilization takes over from a more primitive one, as in the Cape and in Australia: [...] if all the Indians are butchered, a grand extent of country will be gained for the production of cattle: & the vallies of the [Argentine rivers] will be most productive in corn' (*Diary*, 180–1; Barta, 'Mr Darwin's Shooters', 118; cited in Desmond and Moore, 2010, p. 148).

features by which the ancient unity of our species is manifested' and to see unity in diversity (Sachs, 2007, p. 67). Humboldt expresses this view of 'the possibility of common descent' in *Cosmos*:

> Man every where becomes most essentially associated with terrestrial life. It is by these relations that the obscure and much contested problem of the possibility of one common descent enters into the sphere embraced by a general physical cosmography. (Humboldt, 1997, p. 351)

Both Humboldt and Darwin shared an understanding of what Nature's dark secrets meant, both in terms of the workings of Nature and their own personal lives. They not only understood through their travels what was 'distant, foreign, different and abnormal' but also understood this in terms of being on the margins of society. For Humboldt, he was forever on the move like an exile, for Darwin, keeping his ideas secret and avoiding publication for twenty years, a kind of literary exile (Sachs, 2007, p. 63).

For Darwin, this tension between the internal and external forces is expressed in the struggle for existence and the destruction of life. This 'struggle' represents the interrelationship of all the parts that make up Nature whether this is the competition between species or the conditions that are favourable or unfavourable to them, such as drought, flood, different terrains, abundance or shortages of food types necessary for survival. All these factors are interrelated and make up the organic whole of Nature. Humboldt was perhaps the first ecologist (in today's language) in understanding these relationships, but he also had an understanding of Man's aesthetic appreciation of the beauty of Nature as well as his ability to have a destructive influence on it. As discussed earlier, he was amazed by the difference and variety of Nature and its habitats from region to region, but also Man's impact on Nature through agricultural development and Man's impact on Man through slavery. Like Humboldt, Darwin was ahead of his time in recognising the power of *intra-specific* competition (which is a dominant theme in modern ecology that can be expressed in an animal's struggle against drought), as well as *inter-specific* competition (that can be expressed by different species of plant struggling to cover the ground) (Darwin, 1985, p. 116). It is also expressed by the Malthus doctrine in which competing numbers for limited supplies of food affects survival of individuals and species (Darwin, 1985, p. 117).

Darwin sees this destructive element of competition in Nature as quite violent, as if its face has been hit by thousands of 'sharp wedges' again and again (Darwin, 1985, p. 119). But when reading this description, it must be remembered that Nature is attacking itself; it is Nature's internal and external forces that determine natural selection that thereby maintain Nature's own survival. The 'wedges' in a masochistic kind of way help maintain the balance between the species, helping the most successful survive. This also supports the concept of Nature as one reality in which Nature is both creator and product, that is, in Spinoza's terminology, 'nature natured' (Nature as product) and 'nature naturing' (Nature as infinitely productive, infinitely becoming).[14] Darwin had already seen examples of the violent forces of Nature time and time again on his sand walk at Down House past the 'entangled bank', and this contributed to his reflection on the forces of Nature. In addition to his analysis of the data and specimens collected during his voyage on the *Beagle*, Darwin also conducted experiments in his Down House garden. One such experiment was on seedlings showing that the destruction of most was due to predators such as slugs and insects, indicating that the struggle for survival is not always due to the competition between species or the struggle to obtain food (Darwin, 1985, p. 120). In viewing this organic whole of Nature with its interrelated parts, the human mental understanding of it is necessarily included as an integral part just through the process of thinking about it. This can be seen in the use of words that Darwin uses to describe the struggle for survival: 'destroyed', 'enemies', 'choking' and 'prey' (Darwin, 1985, p. 120). All these words indicate a human perspective or interpretation of Nature experienced by Man, representing a personal as well as a scientific narrative of the results of his experiment. According to Darwin, the inner structure of organic beings is determined by their archetypal forms[15] passed down from 'the ancient

---

14 See Richards, 2002, p. 211 and Miller, 2009, p. 106 for their discussion of Spinoza's terms *Natura naturans* and *Natura naturata*.

15 As expressed by Darwin, 'If we suppose that the ancient progenitor, the archetype as it may be called, of all mammals, had its limbs constructed on the existing general pattern, for whatever purpose they served, we can at once perceive the plain signification of the homologous construction of the limbs throughout the whole class' (Darwin, 1985, p. 416). Darwin takes his idea of archetypes from Richard Owen,

progenitor' and the laws of Nature, such as natural selection, and the laws of compensation.[16] As later discussed, these forces or laws can only be seen and understood in the mind's eye through reflection. The sense experiences of Nature do not always appear to provide a direct experience of Nature's 'inner kernel'; the senses or the ideas are experienced as if Nature is providing a veil over its pure Nature making it appear hidden. Yet the mind's eye does provide an insight into the secret forces of Nature if it is seen as organic and made up of the interrelatedness of all other organic beings:

> The structure of every organic being is related, in the most essential yet often hidden manner, to that of all other organic beings, with which it comes into competition for food or residence, or from which it has to escape, or on which it preys. (Darwin, 1985, p. 127)

Darwin's theory of natural selection is undoubtedly his best example of the hidden forces of Nature. His theory banishes the view of the sudden or 'special act of creation'[17] of species. Darwin's forces of Nature come from Nature and act upon Nature, and do not come directly from an external power. This inner law of striving to modify an organic being's structure to gain an advantage does not occur in isolation in Nature and is often achieved in tandem through the mutual support of other organisms. For example, flowers producing the most nectar get visited by the most insects, ensuring their pollination and fertility, and, likewise, insects take advantage of this for their own survival (Darwin, 1985, p. 140). But natural selection does not just stop at producing one organism to provide that advantage. Through the 'physiological division of labour' of producing male and female organs, plants (as well as animals) are able to strengthen this process of produc-

---

citing Owen's 1849 'Nature of Limbs'. However, Darwin objected to Owen's view of a 'continuous operation of creative power' (Darwin, 1985, pp. 58–9).

16 An example of the law of compensation can be found 'in the giraffe, [in which] the head and extremities are developed at the cost of the main body, while in the mole the reverse is the case' (Richards, 2002, p. 447).

17 'If we look at each species as a special act of creation, there is no apparent reason why more varieties should occur in a group having many species, than in one having few' (Darwin, 1985, p. 111).

ing the most efficient outcome of a species' survival. Importantly, Darwin demonstrates how this end result can be achieved through many intermediate steps leading up to this state of perfection (Darwin, 1985, p. 141). Once this state of near perfection has been achieved (through producing the most efficient outcomes), Darwin can see no difficulty in imagining 'humble-bees' adapting their structures to enable them to visit red clover flowers to get the nectar, thus demonstrating the mutual dependency of red clover flowers and 'humble-bees' (Darwin, 1985, p. 142).

In having earlier discussed the concept of organic Nature in terms of the 'web of affinities', it should now be apparent that Darwin's theory of natural selection has its own 'web of affinities' with 'natural selection' forming the centre of its own web. Although Darwin principally uses the concept of the 'web of affinities' to refer to the genealogical relationship between the species, his use of the term can be extended to cover the 'web' of related laws making up his theory of natural selection. It can be argued that the natural law of natural selection, as already discussed, can point to both the past, genealogically in terms of nature's progenitor heritage, and to the future, teleologically in terms of self-improvement[18] (which can be understood reflectively or imaginatively as the 'chain of affinities'). At the same time there are ancillary natural laws that feed into the law of natural selection, as for example with the law of compensation previously mentioned (where the 'laws' are those laws governing the forces of Nature);

---

18   Although 'self-improvement' strictly means no more than 'increasingly effective adaptation to the conditions', Darwin is suggesting that in his view the adaptations can only develop according to predetermined pathways that emanate from the archetypal forms. In this sense, rather like embryonic forms, they can be seen to develop teleologically according to pre-determined rule-governed behaviour. But perhaps in a stronger sense, these predetermined pathways are governed by archetypes that are striving to improve themselves because this is *necessary* for their survival (they have to adapt in order to survive) and in this respect natural selection could be seen to be teleological. Additionally it could also be suggested that the development of the rule-governed archetypes could be interpreted as *necessary* and that therefore the pathways of the archetypes that make up Darwin's 'Tree of Life' are teleological. The ends of each branch end up that way because their structures have predetermined, or programmed them, to do so.

at times they seem to be one and the same, but this is dependent on the degree of closeness Man can get to this knowledge of Nature through his limited direct empirical experience and his reflective experience. For the most part Man only experiences the forces and can only imagine what the laws might be.

To conclude this chapter, it should be emphasized that Humboldt helped Darwin develop a new imaginative way of thinking, enabling him to step outside of his Victorian time warp (if only with one leg). Humboldt enabled him to develop his mind's eye through reflection and imagination. Like Humboldt before him, Darwin was able to step back from the collection of his 'scientific' data and relate them to the context of the impressions made by the sense data on his mind. In the same way as the plant life at different altitudes gave Humboldt an idea of the characteristics of a country's region, so too did the rock strata, petrified trees and location of sea shells give Darwin an idea of the natural history of a region and how life might have developed – it helped spark his imagination and helped him consider the impossible: that things had not always been fixed in time, that they could change.

CHAPTER 3

Darwin's Romantic Theory of Nature

This chapter looks at those German Romantics that can be identified as having had an influence on Humboldt and consequently Darwin. It looks at both Humboldt's work and his contemporaries to determine whether they could be regarded as Romantic and whether their brand of Romanticism informed Darwin's work. The chapter looks at the relationship between literature and science and how they are used to describe the experience of Nature. This gives some background in understanding Humboldt's method of incorporating aesthetic experiences into his interpretation of Nature and how this in turn sparked Darwin's own imagination enabling him to see natural selection in Nature. The chapter also provides evidence to show that Darwin was influenced by Malthus and Townsend in terms of struggle and self-improvement which are important aspects of Darwin's theory of natural selection.

This chapter examines the concept of struggle in terms of what it means for Darwin's theory of natural selection and what it means in terms of an understanding of Nature. Is 'struggle' an example of the force of Nature and is this Romantic? Is it an example of a mental experience reflected through mankind's relationship with Nature? Does it represent a set of objective laws? Or is it a combination of subjective mental experiences and objective laws? In trying to answer these questions in this chapter, the influence of Humboldt's *Personal Narrative* on Darwin needs to be traced back to his voyage on the *Beagle* as this informs his later work such as the *Origin*. Humboldt's organic view of Nature, in which everything is related to everything else, can be seen to help Darwin formulate his theory of natural selection which does not require a Creator to create individual species, as Nature is self-creating – however, this view is not an argument for the non-existence of a God.

In discussing natural selection, the question of self-improvement needs to be raised and the debate about whether this can be seen as teleological needs to be revisited. Are the laws governing Nature necessary and therefore predetermined? Is Nature mechanistic and purposive? Goethe makes the link between the genealogical past and the teleological future and the chapter will examine this in relation to his 'Genetic Method' of thinking about Nature as well as his identification of the unchanging 'inner kernel' that runs through past and present: that is, the internal pattern of a creature remains constant through time, but its outward appearance adapts according to particular environments.

This chapter also traces the concepts of metamorphosis and archetype in Darwin's work as they can be shown to be central to his concept of change as well as the endurance of species. In analysing the concept of Mind in Humboldt and Darwin, reflection and communion with Nature can be seen to help the observer go beyond his own individuality to experience the sublime, and, in so doing, make a contribution to both science and literature that goes beyond the self. Thus through Darwin's imagination and reflection he is able to discover the laws of natural selection.

An important dimension to understanding the 'soul' or power of Nature in terms of its self-generating laws is the ugly side of its conflicts and competitions within and between species, including Man. Darwin was well aware of these conflicts when he went on his walks, noticing plants competing for light and space in which to grow. This was highlighted by his reading of Thomas Malthus which reinforced the view of a complex organic Nature with an inner power both beautiful and ugly, with an attempt by Man to control it, whether through farming, breeding animals and pigeons, managing zoos or botanical gardens, medicine, economics or politics. Malthus' (2008) *An Essay on the Principle of Population* is an example of the latter two. When he wrote it in 1798, it is important to understand that it was written nine years after the first year of the French Revolution in 1789 and eight years after Edmund Burke's speech (9 February 1790) in the House of Commons in which he spoke of his fears of anarchy, violence and atheism, followed by his *Reflections on the Revolution in France* (1790) (Burke, 2010). With the subsequent beheadings in France, the upper classes in Britain feared for their own lands and lives.[1] Against this backdrop, Malthus wanted to demonstrate that unchecked populations continued to expand as they had no internal constraints to control them. Malthus would have been influenced by the debates at the time concerning the *Poor Laws*, and although there is no evidence that he had read Joseph Townsend's *Dissertation* of 1786, his father, Daniel, would certainly have been familiar with it, and so therefore would his son Thomas (Ashley Montagu, 1971, p. 10). In his *Dissertation*, Townsend refers to the Juan Fernández island off the coast of Chile on which the Spanish introduced greyhounds to compete with the goats to help reduce the food supply for British sailors. The result was that the weakest of both species perished with the most 'active' and 'vigorous' surviving. From this, Townsend's following principle can be seen to be the same as expressed in Darwin's *Origin* seventy-three years later:

> The weakest of both species were among the first to pay the debt of nature; the most active and vigorous preserved their lives. It is the quantity of food which regulates the numbers of the human species. (Townsend, 1971, p. 38)

For Townsend, 'hunger' should keep the human species' appetite for reproduction in check, or at least the fear of hunger for his offspring:

> There is an appetite, which is and should be urgent, but which, if left to operate without restraint, would multiply the human species before provision could be made for their support. Some check, some balance is therefore absolutely needful, and hunger is the proper balance; hunger, not as directly felt, or feared by the individual for himself, but as foreseen and feared for his immediate offspring. (Townsend, 1971, p. 44)

Malthus too made the general assumption that the sex drive knows no constraints and that this applies equally to the human race. For the race to survive, humans had to be kept in check by famine, disease or war, or by the intervention/non-intervention of an ethical elite (Hall, 1995). When Darwin read Malthus' *Essay* in 1838 he was living against a similar backdrop of political and social turmoil with similar fears of anarchy and atheism.

1   Hall, (1995).

The *Essay* (and perhaps subconsciously this backdrop) gave him this idea of the struggle for existence which he could see in Nature. It should also be pointed out here that Townsend was widely read in the literature of the Physiocrats, led by François Quesnay and followed by Rousseau and the Romantics as stated earlier. Quesnay was inspired by William Harvey's demonstration of the circulation of the blood and saw economics, as represented by production, exchange and distribution, as a natural law of science maintaining the wealth of the nation in the same way as the heart maintains the health of the individual. Townsend was opposed to the poor laws as he believed they discouraged the poor from working – for him the poor's work in agriculture (on the land) was essential for the creation of the nation's wealth (like the circulation of blood, the poor were seen as an essential cog). As summarized by Ashley Montagu, the Physiocrats' belief was that

> the poor are necessary to co-operate with nature in making its wealth available, by extracting it from the soil so that it can then be converted into consumable form. (Montagu, 1971, p. 7)

This emphasized the holistic, interrelated, organic, Romantic view of Nature, which can also be found in Darwin's works. From the above, we can see that Rousseau (via the other Romantics of the age), in addition to Malthus and Townsend (from the pessimistic standpoint of struggle and competition), had a significant impact on Darwin's moral stance thereby influencing the development of his theory of 'natural selection'.

As demonstrated by the above, the Romantics were relativists in that Nature (including Man) was not fixed, not governed by universal laws, was diverse and forever evolving and improving, as with Darwin's theory of 'transmutation' (now known as 'evolution'). We can already see the seed for this theory being developed when Darwin was on the *Beagle*. Although at this stage he has not yet consciously articulated the question as to whether species are mutable or immutable, he has started the debate by casting doubt on the widely held view at the time that the extinction of species is due to sudden catastrophic events:

> In some countries, we may believe, that a number of species subsequently introduced, by consuming the food of the antecedent races, may have caused their extermination;

> but we can scarcely credit that the armadillo has devoured the food of the immense Megatherium, the capybara of the Toxodon, or the guanaco of the camel-like kind. But granting that all such changes have been small, yet we are so profoundly ignorant concerning the physiological relations, on which the life, and even health (as shown by epidemics) of any existing species depends, that we argue with still less safety about either the life or death of any extinct kind. (Darwin, 1989, p. 164)

This release from mathematical reasoning freed the mind to be creative and imaginative, as we can see from the above example of Darwin using observed data to stretch his imagination to its limits. The imagination becomes the synthesis of reason and feeling, enabling us to reconcile opposites, or what Coleridge calls 'intellectual intuition' (Melani, 2009, p. 2). An example of Darwin's imaginative synthesis to develop his idea of natural selection can be seen in his 'tree' diagram; this demonstrates the transmutation of species from their origins, showing Nature as an interrelated organic whole, not as a machine governed by fixed mechanical laws (Melani, 2009, p. 2). The paradox of the tension between reason and feeling can be seen in Darwin's work on barnacles and worms, ugly perhaps on the outside, but beautiful within when one understands the intricacies of the workings of Nature. Other British naturalists, such as Henry Walter Bates and Alfred Russel Wallace, were also examining these intricacies on the River 'Amazons', experiencing Nature in the raw with dangerous alligators, snakes and, in Wallace's case, malaria. It was during one of Wallace's fevers while thinking about Malthus' *Essay* that he hit upon his version of the theory of 'natural selection' in 1858 and sent it to Darwin (Wallace, 1858, p. 62).[2]

As the above discussion of Malthus' influence on Darwin has shown, struggle in Nature reflects an organic interrelated whole, and one which, through competition and conflict is forever evolving. And it is the contemplation of such a struggle in the 'entangled bank', for example, that enables Darwin to bring reason and feeling together to create his theory of natural selection. In examining Darwin's contemplation in relation to

---

2   Wallace's paper, along with pressure from Huxley, Lyell and Hooker, led to the joint Wallace-Darwin paper presented by Charles Lyell and Joseph Hooker to the Linnean Society of London on 1 July 1858 Darwin, and finally leading to Darwin publishing his own work, the Origin, in 1859 (Wallace, 2008, pp. 6–16).

his insight into Nature, the meanings of experience and Nature need to be further explored as these are related to his concept of Mind. To what extent is Nature objectively experienced and to what extent is it made up of subjective experiences? Wordsworth, for example, sees himself as both an objective observer of Nature describing 'the physical environment around him', and as a 'subjective shaper of experience', mind therefore being seen as 'creator and receiver both'[3] (Nichols, 2004, p. 160). For Goethe on the other hand, Nature can only be experienced through the collective experience of Man through incremental research in infinite time; Nature cannot be viewed at any single point in time (Hadot, 2006, p. 179) as the status of scientific knowledge only represents that moment. New elements are forever being 'discovered', which in turn rewrite earlier states of 'knowledge', and this expansion of our understanding never stops and is therefore infinite. The difficulty with the concept 'Nature' is that it can have very many different meanings: do we mean individual natures corresponding to distinct species or variations each with their own forces, or do we mean one Nature with one force or set of laws flowing through all the species? Are all these natures or species equal on the same level or are they arranged hierarchically? These are issues that need to be examined alongside the questions, 'Is there any evidence to show that Darwin viewed Nature as objectified on the one hand, or divine on the other?' and 'were these views influenced by the Romantics and Humboldt in particular?'.

The above questions will now be looked at in relation to Humboldt and Darwin's 'union of poetry and science' as this provides a clearer understanding of what is meant by the experience of Nature. As already discussed, Darwin modelled his work on Humboldt's *Personal Narrative* of his expedition through the rainforests of the Orinoco and the River Negro and states this debt of gratitude in his autobiography (written in 1876). From this he learnt the art of accurate and vivid descriptions of his observations:

> During my last year at Cambridge, I read with care and profound interest Humboldt's *Personal Narrative*. This work, and Sir J. Herschel's *Introduction to the Study of Natural*

---

[3] This is best exemplified in Wordsworth's poem, 'Tintern Abbey'.

> *Philosophy*, stirred up in me a burning zeal ... No one or a dozen other books influenced me nearly so much as these two. (Darwin, C. 1995, p. 23)

This influence is apparent when reading Darwin's diary of his voyage:

> I believe from what I have seen Humboldt's glorious descriptions are & ever will be forever unparalleled: but even he with his dark blue skies & the rare union of poetry with science which he so strongly displays when writing on tropical scenery, with all this falls far short of the truth. The delight one experiences in such times bewilders the mind,-if the eye attempts to follow the flight of a gaudy butter-fly, it is arrested by some strange tree or fruit; if watching an insect one forgets it in the stranger flower it is crawling over, – if turning to admire the splendor of the scenery, the individual character of the foreground fixes the attention. The mind is a chaos of delight, out of which a world of future & more quiet pleasure will arise. – I am at present fit only to read Humboldt; he like another Sun illuminates everything I behold. – (Keynes, R. D. 1988, p. 69)

This could be interpreted as an example of 'the rare union of poetry and science' in Darwin's own writing inspired by Humboldt. Is this also an example of the Romantic union between the subject (the beholder) and the object of Nature (that which is observed), or is it merely an enthusiastic description of Nature? Can the two be distinguished and how would one go about distinguishing them? The blurring between the two makes this difficult. There are hints of this with the use of the words 'bewilder' and 'chaos' expressing a mass of sensations that the subject has difficulty in organizing, yet at the same time a feeling of delight that could be seen as an internal sensation within Nature itself. But like any scientist dealing with a mass of data, it has to be organized and classified in order to make sense of it, and this is perhaps what Darwin is alluding to when he refers to 'a world of future & more quiet pleasures will arise'. This could be an example of Humboldt's Romanticism in poetry and science – the ability to capture the sensations of a holistic Nature where the edges of subjectivism and objectivism blur, yet, at the same time, being able to make sense of them scientifically through classification and analysis but without diminishing the beauty of the butterfly.

As discussed earlier, Humboldt combined his emotions and personal experience with his data gathering when describing Nature. As a Romantic, he believed that the rational scientific mind could be complemented by an aesthetic judgment able to penetrate 'the deep structures of reality, a path overlooked by most Enlightenment thinkers' (Richards, 2002, p. xvii). The Romantics attacked the mechanistic view of science, replacing it with an 'organic' interpretation in which their conceptions of Nature were not just formulated from abstract numbers but were created from their immediate scientific experiences and intimate relationships inspired by love and hate (Richards, 2002, p. xviii). Richards argues against the view that poets, artists and scientists can 'formulate conceptions unmarked by their own lived experience' [and that the mind can] 'float free, detached from the imperfect life that produced it' (Richards, 2002, p. xviii).

Gillian Beer also sees Darwin as an example of a 'Romantic materialist' in which the instruments of science through speculation complement the mystery of Nature (Beer, 1983, p. 252). The observation of Nature is at the heart of any discussion of whether Darwin was a Romantic. What is observed and captured through the senses may be material but the pleasure of the experience is Romantic, whether optimistic or pessimistic:

> Observation is charged with sensory power. In both [Darwin and Hardy] the material world is described simultaneously in terms which may lend themselves to an optimistic or pessimistic interpretation, but which function as terms through the pleasures of observation. (Beer, pp. 244–5)

Here the material instruments of science enable the scientist to look towards the future without destroying the mysterious qualities of Nature. Paley's Creator or Deity as the prime cause of Nature may have been removed, but it can be argued that the materialism of science has not removed its mystery, its 'Romanticism', but has added to it. Although the concept 'Romantic materialism' seems to be an oxymoron, the terms can be seen to complement each other. Although mind is not seen as existing independently of matter as in Cartesian dualism, the spiritual still exists in terms of defining the experience of the forces of Nature even if the causes are not known. The material world can therefore be seen as Romantic.

Darwin uses many metaphors in his texts and these reflect Beer's argument above that the pleasures of observation, whether optimistic or pessimistic, are Romantic. The most common metaphor used is 'struggle'. On one level it depicts plants and animals fighting for survival both as individuals and as species. The consequent violent destruction of individuals through this struggle is represented by his use of the 'wedge' metaphor[4] (discussed earlier) and is an extreme example of pessimism. Although the notion of the 'entangled bank' (that is, species of vegetation competing for light, nutrients and space) expresses the pessimistic struggles of individuals and species, it also expresses the optimistic pleasure of observing the varied forms of life that reproduce themselves, adapt to their environment and to each other in living in mutual harmony. The concept of 'struggle' in Darwin, the man and his works, is multi-layered. In Darwin's theory of natural selection he is grappling with the struggle in Nature between individuals within and between species, all struggling to survive. It could be argued that in his writing he is struggling to express his own meaning through the existing Victorian creationist vocabulary of 'select', 'create', and 'design'. Beer, for example, argues that 'his description is necessarily conditioned by the assumptions and beliefs condensed in the various kinds of discourse active at the time he was writing' (Beer, 1983, p. 51). Additionally she states that as 'natural history was still imbued with natural theology, [...] Darwin was therefore obliged to dramatize his struggle with natural theological assumptions within a language weighted towards natural theology [and that therefore] he must write against the grain of his discourse' (Beer, 1983, p. 53).

As a writer and scientist, Darwin struggled with his conscience to publish the *Origin* knowing the concerns his views would raise within the church.[5] For example, how could Man be made in the image of God if he shared a common progenitor with primates, and where would that leave morality if it did not emanate from a transcendent Being? Darwin's view in *The Descent of Man* that Man's morality developed naturally only added to the fear that if Man were no different from animals he may behave like

---

4   Darwin, 1985, p. 119.
5   Robert Chambers' anonymous publication in 1844 of his *Vestiges of the Natural History of Creation* had already created a storm by putting forward the view of the transmutation of species (bringing together the views of other naturalists at the time) (Chambers, 2010, p. 64).

them. Having design come from within Nature rather than from a Divine being made his work seem atheistic. Also Darwin's tree diagram used to illustrate the development of species could be seen as demolishing the idea of an unfolding Divine plan (Brooke, 2003, pp. 192–5). Could Darwin's 'struggles' be regarded as Romantic? Some of the struggles could be seen as the consequences of being a scientist in Victorian England at a time of change. But they could also be seen as a consequence of his Romantic way of experiencing and interpreting Nature. To what extent were his own personal struggles a reflection of an awareness of himself as an active participant in the 'struggle' of Nature? – for example, is there evidence to show he was aware of the problems of ill health associated with in-breeding in humans as well as animals? If so this would be another example of Beer's 'pessimism' forming part of Darwin's 'Romantic materialism'. Likewise, is there any evidence to show that this Romantic 'pessimism' in Darwin's life is offset by Romantic 'optimism'? T. E. Hulme (Hulme, 1911, pp. 1–3) believes there is (although he objects to Romantic verse). He argues that Darwin's theory, as with the Romantics, represents a progressive view of nature. This could be seen as entailing optimism as it 'regards man as a well, a reservoir full of possibilities', evolving and adapting incrementally (and is therefore 'intrinsically good').

As already defined, Nature can be seen as Romantic if its structure is seen as consisting of an archetypal organic core, and having a relationship to Mind (the 'imagination') with aesthetic and moral features. These aspects of Romanticism will now be analysed to see what influences they may have had on Darwin and his work. Darwin's example of the possibility of an insect-eating bear (Darwin, 1985, p. 215),[6] taking to water and developing into a whale, is very similar to Goethe's example of the descent of the modern sloth. Goethe[7] supposed the giant sloth had first been a whale that got trapped along a swampy, sandy beach and to bear its weight developed large limbs passing these on to its descendants (Richards, 2002, p. 485). Both these examples from Darwin and Goethe can be seen as Romantic examples of Man being part of Nature through his ability to stretch his scientific imagination (his Romantic mind) through argument by analogy.

6   Referred to earlier.
7   In Goethe (1985–1998, 12:246–7), cited in Richards, 2002, p. 485 footnote 218.

This is not the creation of a fanciful world but one created out of perceptions being related logically through a mentally constructed history of Nature; a timeline constructed from the evidence, for example, of rock formations, present species' behaviours and characteristics. Although scientific, these hypotheses could also be seen as aesthetic due to the way Nature impresses itself on Man's senses and how Man responds to these perceptions, how he experiences the interrelationships of, for example, the sea, the sky, the mountains and the forests – the poetic context of the experience adds to the scientific. The whole experience is objective yet subjective, public yet personal.

This holistic view of Nature combining the subjective with the objective through the Romantic imagination can be seen in the Romantic idea of the eye. If the intricate eye of Man or any animal is not seen as the design of a Creator but as the product of Nature as producer (from its own archetypes) perfected through development over time, then this could be seen as Romantic (and this is a Romantic idea coming from Man's Romantic imagination through his experience of Nature). It could be a Romantic idea because Nature could be seen as the designer; but not in the sense of Nature having the idea of the perfect eye as one design and then creating it. From the first development of a simple light-sensitive cell, the eye is developed over hundreds of millions of years, going through many, many stages, each stage fitted for its function at that time until the present stage is reached. The development of the eye, as with all other organs, could be seen as Romantic because it has developed organically; organically in the sense that its development is due to its interrelationships and interactions within Nature (for example between the eye and other species and habitats) as well as within the eye's owner (for example between the eye, the brain, the heart).

Darwin was very concerned that the creation of the eye could be used as an argument against his theory of natural selection and he admitted that initially it did seem absurd to imagine that this could have developed in this way. Whilst at Cambridge, Darwin had read Paley's account[8] of the intricate creation of the eye as an argument for a divine Designer and this seemed difficult to argue against. For Paley the eye was an example of 'crea-

---

8   William Paley's *Natural Theology or Evidences of the Existence and Attributes of the Deity* (1802).

tive intelligence' constituting 'the order and beauty of the universe' (Paley, 2005, p. 26). However, Darwin then went on to argue that this did not seem to be so absurd if one imagined it being developed slowly over a very long time span in which each slight development had been of use to its owner:

> To suppose that the eye, with all its inimitable contrivances for adjusting the focus to different distances, for admitting different amounts of light, and for the correction of spherical and chromatic aberration, could have been formed by natural selection, seems, I freely confess, absurd in the highest possible degree. Yet reason tells me, that if numerous gradations from a perfect and complex eye to one very imperfect and simple, each grade being useful to its possessor, can be shown to exist; if further, the eye does vary ever so slightly, and the variations be inherited, which is certainly the case; and if any variation or modification in the organ be ever useful to an animal under changing conditions of life, then the difficulty of believing that a perfect and complex eye could be formed by natural selection, though insuperable by our imagination, can hardly be considered real. (Darwin, 1985, p. 217)

For Darwin the development of the eye is an important example against the Creationist view, as the present state of the eye is the result of many, many intricate steps of development over a period of time in which all the steps collectively make up the whole eye. All the steps are interdependent along the chain making up the present eye. The eye is its historical makeup and is therefore organic. So too with all other organs and creatures. All their structures both past and present reveal some purpose and these reflect the laws of growth that are part of the principle of natural selection (Darwin, 1985, p. 228). The evolutionary development of the intellect in Man, for example, can also be seen to be Romantic in the sense that it is part of Man's organic reflection[9] on the perception of the relationships in Nature such as that between fossils, rock formations and time lines of species as well as the aesthetic appreciation of beauty in Nature.

---

9   This intellectual (or imaginative) reflection on Nature is organic due to all the interrelationships and interactions that make this possible: sense data received through the eyes, the ears, touch, taste, smell and their processing by the brain; the further intellectual processing of these experiences and emotional interpretations, and comparisons made between experiences and the experiences of others both in present and historic time.

For Romantic naturalists such as Humboldt this constant flux of life and death, of constant change, of constant development, highlighted the tension between the finite and the infinite; the infinite representing all finite beings, all change. For both Humboldt and Darwin, experiencing the volcanoes, earthquakes and jungles of South America was an experience of the sublime,[10] filled with trepidation in which their feeling of the infinite power of Nature was tempered by their own insignificance as mortal human beings. As scientists (in the modern sense of the word) they were able to describe and measure Nature with their elaborate scientific equipment, yet were unable to totally control it (despite Man's commercial exploitation of forests, plants and plantations, trade routes and slavery).

This experience of Nature of which Man is a part is self-contained. Nature creates itself from its own archetypes and is not created by some external Creator. This is the 'One Reality Nature'. The idea of there being one reality goes back to Spinoza. He believed that the whole of Nature could be regarded as Divine and that God was therefore not separate from Nature but *was* Nature; all the parts of Nature made up the whole which was God.[11] There was no external creator. Nature created itself. It was both creator and product. As Spinoza put it, nature was *natura naturata* or 'nature natured' (that is, it was the product) but also nature was *natura naturans* or 'nature naturing' (that is, it was infinitely productive, infinitely becoming).[12] Individuals who experienced this whole of Nature were able to experience the intellectual love of God which was an understanding beyond reason (Richards, 2002, p. 211). This wholeness of Nature was taken up by Schelling who believed that the forms in Nature were the same as in the self (that is, they were isomorphic). A genius who was able to achieve this

---

10   This is the quality of greatness, whether physical or spiritual which is beyond all possibility of measurement.
11   According to Clement Carlyon (who met Coleridge in Germany in 1799), Coleridge was influenced by a Priestleyian interpretation of Spinoza's materialism, in which our own thought should be seen in terms of the properties and powers of matter (Halmi, 2012, unnumbered page, l.l. 1–31).
12   Spinoza's identification of God with Nature was explained in his posthumously published *Ethica Ordine Geometrico Demonstrata* in 1677.

intellectual understanding of the whole of Nature would therefore meet his double.[13] Schelling believed that this way of thinking about Nature was a necessary therapy to counteract Man's separation between the self (ideas or perceptions) and the world (objects or things) (Richards, 2002, p. 136).

Through the adventurous explorations, experiences and scientific works of Humboldt and Darwin, this separation between the self and the world is removed. Both Humboldt's *Personal Narrative* and Darwin's *Voyage of the Beagle* (also a personal narrative) show the explorer immersing himself in the experience of Nature both as an individual and as a scientist through measurement, data collection and analysis. In both, they narrow the division between their ideas and the objects of their perceptions. Both experienced the incredible diversity of Nature and its corresponding habitats as well as the personal dangers and illnesses associated with the rawness and closeness of Nature. Darwin was able to take one step back and observe a greater wholeness of Nature through having more data to hand (through more knowledge being made available to him from other naturalists as well as his own additional data and experiences). Also, through his understanding of archetypes, Darwin was able to apply his knowledge genealogically to evidences of the origins of Nature (for example, fossils and similarities between species). Both Humboldt and Darwin had perceived Nature aesthetically and intellectually, feeling at one with Nature through their experiences, but Darwin was one step ahead in that he was able to schematize it through his theory of natural selection, capturing the essence of the secrets of Nature.

The above notion that Nature can be both objective and subjective, or material and spiritual, posits the view that Nature can have self-contained structures and laws such as Darwin's natural selection. But to be free, Nature had to come from organic self-consciousness. This self-consciousness can be seen in Darwin's conceptual reflections on Nature, in which his overview of the history of Nature enables him to relate geological changes to changes in species' development; in which each flower's petal, each bird's beak, each tree's leaf, each mountain's top and fossil shares an historical

---

13   Richards, 2002, p. 116, cites Humboldt as an example of this.

relationship. The necessity-freedom paradox can be seen in Darwin's genealogical 'tree of life' in which the archetypal necessary forms run through the species, and the freedom to adapt and change run through the variations/incipient species or intermediate species (yet conforming to necessary laws on how to do this). The necessity-freedom paradox is a Romantic interpretation of Nature as it treats it as an historical vein running through all beings bound by the laws of Nature yet free to develop and change. As reflective beings, humans are able to experience this over and above the passive impressions of sense data. Through reflection humans can experience their own part in Nature's historical development, through its flux, its laws, and through a glimpse of its secrets.

Darwin's genealogical classification through his 'tree of life'[14] gives a true plan of creation through his own self-reflective consciousness based on personal empirical experience of Nature but also the collective understanding based on his own and other naturalists' reflections and concepts. Man's understanding of the history of Nature can be seen as a mental reconstruction of its organic whole: its laws, its genealogy, its relationships (between laws and laws, between species and species, between individuals and individuals, between past and present). This can be regarded as the face of Nature's self-consciousness if Man's Mind through his scientific and aesthetic 'reflective' understanding is seen as reflecting this organic whole. For without any experience or reflection of Nature there is no Nature. Our human knowledge and understanding of the world only exists as a result of our experience of it (unless one believes in Kant's 'noumena', or thing-in-itself,[15] existing independent of human consciousness). Collectively, and therefore organically, the history of Nature provides a science of the origin of species and this is achieved through its fossil and geological remains,

---

14   His 'tree of life' diagram in Darwin, 1985, pp. 160–1.
15   Kant defines the 'thing-in-itself' as follows: 'The existence of things, that which appears, is not destroyed as in real idealism; rather it is only shown that we cannot know anything about them, insofar as they are things in themselves, through the senses' (Kant, *1957*, 6:153 (A 64). Cited and translated by Richards, 2002, p. 63). We only experience the *appearance* of the *noumena* through our senses without experiencing the *noumena* itself, as, according to this view, it is forever hidden.

its revelations of causes and laws of growth and variation, as well as the concepts of migration and climate change:

> When we regard every production of nature as one which has had a history; when we contemplate every complex structure and instinct as the summing up of many contrivances [...]. A grand and almost untrodden field of enquiry will be opened, on the causes and laws of variation, on correlation of growth, on the effects of use and disuse, on the direct action of external conditions [...]. Our classifications will come to be, as far as they can be so made, genealogies; and will then truly give what may be called the plan of creation. (Darwin, 1985, p. 456)

Once again the 'history' of Nature is key here, in which modifications can help the scientist understand the 'laws of variation'; in which this genealogical view of Nature can be used to classify its species not in isolation but according to their historical relationships.

As a Romantic, Schelling's 'Dynamic Evolution'[16] argued that Nature achieved its absolute state through self-conscious expression (creative mind), that Nature was never at rest and was always seeking perfection. Darwin could also be read in this vein if the perfection aimed at by species is seen as the struggle to achieve the best form in life to achieve an advantage over other species. Although teleologically nothing specific may be sought, the seeking of perfection produces the same result, creating the illusion that evolution is teleological. For Schelling, the organic unity of Nature was achieved teleologically as opposed to genealogically (Richards, 2002, p. 297). Superficially this might seem to indicate that Darwin's theories were opposed to the Romantics as his theory of Nature was principally genealogical, that is, facing backwards in time towards a primordial progenitor (*vis a tergo*) rather than facing forwards towards

---

16 Schelling defined 'Dynamic Evolution' as follows: 'Since nature must be thought as infinite productivity [*unendliche Produktivität*], conceived really as occurring in an unending evolution [*in unendlicher Evolution*], so its fixity [*Bestehen*], the resting place constituting the natural product (of the organic, for example) must be represented not as an absolute rest, but only as an evolution continuing with an infinitely diminishing rapidity or with an infinite retardation' (in Schelling, 1927b, 2:287 [III:287]). Cited and translated by Richards, 2002, p. 144, footnote 67.

an archetypal attracting force (*vis a fronte*).[17] Yet there does seem to be a telic element in Darwin's theory when it comes to his genealogical 'tree of life', since although it is stretching backwards in real time, enabling the identification of past progenitors, it is also identifying timeless archetypal forms in Nature that exist beyond the original progenitors that existed in a particular point in time. These archetypal forms seem to express a timeless force in Nature, pulling species towards a greater form of perfection, and it is these forces of Nature and archetypes that are expressed in Darwin's theory of natural selection.

---

17 'Darwin presumed that organic unity could be explained by an efficient *vis a tergo*. Schelling rather conceived it as arising teleologically, as a *vis a fronte*, as it were' (Richards, 2002, p. 305).

CHAPTER 4

Darwin's Romantic Theory of Mind

This chapter examines Darwin's theory of Mind, how it is Romantic and how he has been influenced by other Romantic thinkers such as Goethe. Darwin's own thought processes were shaped by the world he saw, both through empirical perception and through intellectual reflection. This mental reflective insight into the working laws and forces of Nature that became his theory of natural selection is akin to Goethe's vision of the vertebra archetype referred to on his second Italian Journey in 1790, in his *Zur Morphologie* (1823):

> [...] as I lifted a battered sheep's skull from the dune-like sands of the Jewish cemetery in Venice, [...] I immediately perceived the facial bones were likewise traced to the vertebrae. (Translated and cited by Richards, 2002, pp. 491–2)

This reflective insight is also referred to in Goethe's discovery of the plant archetype (the 'Primal Plant' or *Urpflanze*) during his first Italian journey in a letter to his friend Johann Herder on 17 May 1787:

> The Primal Plant is going to be the strangest creature in the world [...]. With this model and the key to it, it will be possible to go on for ever inventing plants and know that their existence is logical; that is to say, if they do not actually exist, they could, for they are not the shadowy phantoms of a vain imagination, but possess an inner necessity and truth. The same law will be applicable to all other living organisms. (Goethe, 1970, pp. 310–11)

There are various concepts here taken up by Darwin. The concept of 'primal' or first form is not an archetypal form representing one individual or species at any point in time replaced by others. It is an *underlying structure* that gives life to present and all future forms forever reproducing themselves (and in Darwin's case, forever modifying themselves). This is what Goethe means by the possibility of this form 'for ever inventing plants' and 'if they do not actually exist, they could'. These mental ideas or forms are not merely ideas but are ideas governed by laws and are therefore not merely 'shadowy phantoms of a vain imagination, but possess an inner necessity and truth'. Goethe provides natural historians such as Humboldt and Darwin with a 'model and the key' to unlock the secrets of Nature. In this sense the necessity makes the laws of Nature governing our experiences 'rational', yet as discussed earlier, through their organic nature and the nature of one reality, 'empirical'.

Darwin's indebtedness to Goethe's Romantic notion of archetypes can be seen in one of his two early essays (1842–1844) where he echoes the same theory of morphology:

> There is another allied or rather almost identical class of facts admitted under the name of Morphology. These facts show that in an individual organic being, several of its organs consist of some other organ metamorphosed: thus the sepals, petals, stamens, pistils, &c. of every plant can be shown to be metamorphosed leaves [...]. The skulls, again, of the Vertebra are composed of three metamorphosed vertebrae, and thus we can see a meaning in the number and strange complication of the bony case of the brain. (Darwin, 1909, p. 215)[1]

When Darwin views 'an individual organic being', he sees 'several of its organs consist[ing] of some other organ metamorphosed' and this is similar to Goethe's 'Primal Plant' or 'leaf' archetype. And in Darwin's 'metamorphosed vertebrae' can be seen Goethe's sheep's skull.

As the above shows, reflection is a part of the process of becoming and development of the individual in Nature. Through consciousness the self develops an understanding of the objective world but through the individual's experience also becomes part of the individual's subjective world. This view is expressed by Schelling who believes that self-consciousness and the world are one:

> With the first consciousness of an external world, there also arises the consciousness of my self; and conversely, with the first moment of my self-consciousness, the real world appears before me. The belief in a reality outside of me is established and grows with the belief in the existence of my self; one is as necessary as the other. Both – not speculatively separated but in their complete and intimate interaction – are elements of my life and all my activities. (From Schelling's *Werk* (1797), cited by Richards, 2002, p. 133)

---

1   Also cited in Richards, 2002, p. 435, footnote 77.

For Schelling 'with the first moment of my self-consciousness, the real world appears before me'. The belief in both the self and the outside world are necessary for the life of an individual. Both are intimately connected through their interaction, and this is a reflection of the concept of 'one reality' and 'organicism' discussed earlier.

In the same way as Schelling views the self as forever becoming through experience, so too does Darwin view the world of Nature's species as forever striving to develop themselves through modifications, forever seeking advantage. The argument for an organic nature of Mind can therefore be taken to another level. Not only are the species in Nature striving for development, and not only can they be seen to be made up of a collective Mind in the form of archetypal representations, but they also play an important part in Man's development of scientific understanding through reflection.

The empirical self (referred to above), forever becoming through empirical experiences in time and place, can be seen in Humboldt's aesthetic descriptions of Nature. For Humboldt (as discussed earlier), Nature is much more than a collection of scientific data. It is an interaction between empirical experience and the Mind and the way Mind reacts to Nature. The subjective mental experience is also a form of data and should be valued scientifically. The mental reactions are like the colours on an artist's palette used to paint a landscape and are just as important as scientific measurements taken by instruments:

> The Pico de Teide is not situated in the Tropics, but the dryness of the air, which rises continuously above the neighbouring African plains and is rapidly blown over by the eastern winds, gives the atmosphere of the Canary Islands a transparency which not only surpasses that of the air around Naples and Sicily, but also of the air around Quito and Peru. This transparency may be one of the main reasons for the beauty of tropical scenery; it heightens the splendours of the vegetation's colouring, and contributes to the magical effects of its harmonies and contrasts. If the light tires the eyes during part of the day, the inhabitant of these southern regions has his compensation in a moral enjoyment, for a lucid clarity of mind corresponds to the surrounding transparency of the air. (Humboldt, 1995, p. 36)

Scientific measuring instruments can only go so far in describing the physical terrain. The instruments can measure the humidity, the strength of the wind and the amount of light. But it is the combination of these physical entities and how they affect the mind that creates the atmosphere and makes the region distinguishable from other regions. It is the 'transparency' which is mentally experienced and which gives the area its 'beauty' and 'the magical effects of its harmonies and contrasts'. In fact Humboldt even goes further than this: the mental experience is more than just an objective picture made up of empirical data; it is a subjective experience that goes to the core of the self but one that can be shared by others as it gives 'moral enjoyment'. And this experience is forever reciprocal as it is constantly flowing between the individual and Nature and back again between Nature and the individual. The 'clarity' produced by the experience is mirrored by the clarity of mind, which in turn is able to give the 'inhabitant' a clearer experience of Nature. But these mental images are more than mere subjective reflections; Humboldt is also intent upon sharing his experiences and knowledge with his readers, and, in so doing, is drawing the readers' own eyes to focus on the truths of Nature, in particular our own finite nature within the infinity of Nature (Sachs, 2007, pp. 47–8):

> The earnest and solemn thoughts awakened by a communion with Nature intuitively arise from a presentiment of the order and harmony pervading the whole universe, and from the contrast we draw between the narrow limits of our own existence and the image of infinity revealed on every side, whether we look upward to the starry vault of heaven, scan the far-stretching plain before us, or seek to trace the dim horizon across the vast expanse of ocean. (Humboldt, 1997, p. 25)

This 'communion with Nature' creates a feeling of paradox since it enables us to see our smallness, our finiteness, in relation to the vast infinity of Nature, yet we are also part of Nature in terms of our existence in present time and our heritage in past time. This whole experience creates a mixture of familiarity and strangeness (Sachs, 2007, p. 47), the same feeling of ecstasy and dizziness that Darwin experiences when he first arrives at the Tropics. Humboldt feels particularly disorientated when leaving the Canaries behind and entering the 'torrid zone'[2] as the nightly observations of the stars begin to show completely different patterns of what he is used to and he has to rethink his celestial navigation:

---

2   Hot, dry, parched belt of earth between the Tropic of Cancer and the Tropic of Capricorn.

> A strange, completely unknown feeling is awoken in us when nearing the equator and crossing from one hemisphere to another; the stars we have known since infancy begin to vanish. Nothing strikes the traveller more completely about the immense distances that separate him from home than the look of a new sky. The grouping of great stars, some scattered nebulae that rival the Milky Way in splendour, and regions that stand out because of their intense blackness, give the southern sky its unique characteristics. This sight strikes the imagination of those who even, without knowledge of the exact sciences, like to stare at the heavens as if admiring a lovely country scene, or a majestic site. You do not have to be a botanist to recognize immediately the torrid zone by its vegetation. Even those with no inkling of astronomy know they are no longer in Europe when they see the enormous constellation of the Ship or the brilliant Clouds of Magellan rise in the night sky. Everything on earth and in the sky in the tropical countries takes on an exotic note. (Humboldt, 1995, pp. 41–2)

The openness of Humboldt's 'imagination' enables him to 'see' beyond his immediate surroundings. He is able to understand the changes in vegetation that give each region its characteristics, not only in terms of its position on earth but also in relation to the stars and constellations. Here is a paradox which probably contributes to his disorientation. On the one hand Man's view of the universe is subjective, giving him experiences which spark off his imagination. It is this egocentric view of the universe that gives Man his identity, which creates the 'I' that experiences and interprets all the sensations he receives. Yet as soon as Man moves to a different part of the world and experiences it in a different way, the solidity of his identity seems to crumble. It is as if the laws of Nature themselves have changed and that which was previously certain is no longer certain. Yet there is no fear here, just wonder and the need to adapt to this new environment.

Humboldt also sees Nature differently as he steps into the South American jungle and sees both the diversity and the lack of space for plants to develop, which Darwin echoes when he experiences the same scenes. He is again 'shocked' by what he sees and 'confused' as to what is leaf or fruit (Humboldt, 1995, pp. 83–4). On 7 February 1800 Humboldt travelled from Caracas to the interior as if in a time machine. His journey took him through space and time, seeing inhabitants in the coastal towns living in the eighteenth century and those in the interior living in ancient times as savage Indians (Botting, 1973, pp. 93–4). This strangeness of diversity in different stages of development opened up his mind to the concept of development in time and place, the very concept of natural history itself.

This disorienting mixture of familiarity and strangeness creates an openness of mind that enables the observer, through his communion with Nature, to learn from Nature. Instead of the explorer penetrating Nature to exploit it for his own self-interested purposes, Humboldt is opening himself up to be penetrated by Nature in order to learn from its innermost secrets and truths (Sachs, 2007, p. 60). Becoming disorientated through the experience of strangeness and difference (or diversity) helps the observer to learn to 'see' Nature. It enables the observer to break through the filters imposed on him through his own culture and by questioning them is able to develop a fresh understanding of the world. Disorientation through travel enables this opportunity. This way of looking at the world was one that Humboldt learnt from his mentor George Forster (Sachs, 2007, p. 57). The comingling of the familiar and the strange can be seen in Humboldt's experience of the trees that produce sap-like milk,[3] the man who was able to feed his motherless child with milk from his own nipples[4] and the Ottamac Indians who 'swallow a prodigious quantity of earth' to prevent the feeling of hunger.[5] In this 'new world [...] of untamed and savage nature' (Humboldt, 1995, p. 178), Humboldt finds that Nature speaks to the soul (Sachs, 2007, p. 61). In allowing Nature to overcome the traveller, he does not impose his

---

3   'What moved me so deeply was not the [...] jungles, nor the [...] rivers, nor the mountains [...], but a few drops of a vegetable juice that brings to mind all the power and fertility of nature. On a barren rocky wall grows a tree with dry leathery leaves; [...]; the branches appear dry, dead. But if you perforate the trunk, especially at dawn, a sweet nutritious milk pours out' (Humboldt, 1995, p. 158).

4   'Francisco Lozano, a labourer [...] breast-fed a child with his own milk. When the mother fell ill, the father [...] took it to bed and pressed it to his nipples [...]. The irritation of the nipple sucked by the child caused liquid to accumulate [...]. The father [...] suckled his child two or three times a day for five months [...].' (Humboldt, 1995, p. 87).

5   'The Otomacs eat earth; every day for several months they swallow quantities of earth to appease their hunger without any ill effect on their health [...]. The following are the true facts, which I verified. The Otomacs, over months, eat three quarters of a pound of slightly baked clay daily [...]. It is clear that the sensation of a full stomach came from the clay, and not from whatever else they might eat' (Humboldt, 1995, pp. 266–7).

expectations onto Nature and is therefore better able to be inspired to describe it more accurately (Sachs, 2007, p. 95). As expressed by Humboldt:

> [I have always] considered Nature in a two-fold point of view. In the first place, I have endeavoured to present her in the pure objectiveness of external phenomena; and, secondly, as the reflection of the image impressed by the senses upon the inner man, that is, upon his ideas and feelings.[6] (Cited by Sachs, 2007, p. 43)

Here Humboldt combines 'pure objectiveness' with 'his ideas and feelings'. For him this is 'Nature'.

Earlier reference was made to Beer's observation that for Darwin, what is observed and captured through the senses may be material but the pleasure of the experience is Romantic. So too with Welpley's[7] analysis of *Cosmos*, in which the contemplation of Nature creates 'sensuous' enjoyment (or imagination) and 'intelligent' enjoyment. The reader of *Cosmos* sees the magic and wonder of Nature, its interconnectedness and thereby his own connection to Nature (like Darwin's 'web of affinities'). In this unity of diversity all are equal, and the printed text of *Cosmos*, which makes this awareness possible, shows how art and science can complement each other. Schiller[8] was critical of Humboldt for dissecting Nature rather than revering it as a God. But as Welpley has demonstrated, Humboldt can be seen as a combined Romantic poet and empirical scientist through his use of 'imagination' and 'understanding'. Schelling was also critical of Humboldt for using a mathematical system of shorthand and symbols ('pasigraphy') to describe Nature, but Humboldt believed that knowledge could only come from verifiable experience which satisfied both the mind and imagination (McCrory, 2010, p. 50). Being engaged with

---

6 From Humboldt, 1849–1870, 3:I.
7 His analysis was published with the Latin epigraph 'Mens ingenti scientiarum flumine inundata', translated as 'A mind inundated with a great flood of sciences', in Welpley, 1846, pp. 598 and 602–3 (cited by Sachs, 2007, p. 396, footnote 58).
8 Humboldt contributed an essay, 'The Life Force, or the Genius of Rhodes', (a biochemical hypothesis in the form of a poetical allegory) to the poet Schiller's philosophical journal, *Die Horen* (The Hours). Schiller saw Humboldt as merely reducing Nature to symbols and facts (Botting, 1973, pp. 339–40).

Nature enables the soul to be penetrated by it, passing on knowledge to the observer. Paradoxically this is a union in Nature in which the subjective and objective become one, and can be seen in Humboldt's description of his favourite creature on Tenerife:

> But the bird from the Canary Islands that has the most agreeable song is unknown in Europe. It is the *capirote*, which has never been tamed, so much does he love his freedom. I have enjoyed his sweet and melodious warbling in a garden in Oratava, but have never seen him close enough to judge what family he belongs to. (Humboldt, 1995, p. 37)

It is for this reason that in *Cosmos* he disagrees with Burke that increased knowledge can ruin a person's experience of Nature[9] (Sachs, 2007, p. 48). For Humboldt, knowledge expands the imagination and enables the individual to 'see' the structure of the world; like Goethe, this can be seen as 'a microcosm in a leaf'.[10] This description of the bird that so much loves its freedom could also be seen as referring to Humboldt's desire to free Man's soul through the gaining of knowledge of Nature, but not just through the gaining of scientific data but through the sensuous delight in interacting with Nature, through the enjoyment of the 'sweet and melodious warbling' that lifts the heart through the experience of Man's oneness with Nature (and this is another example of what Beer calls 'the pleasure of the experience', making it more than something physically captured by the senses).

This 'Humboldtian Method' of combining 'understanding' and 'imagination' (scientific analysis and aesthetic impression), can be seen in Darwin's *Origin*. Here there are examples of intellectual reflection in his analysis of geology in support of his theory of natural selection. One of the main

---

9   'I cannot, therefore, agree with Burke, when he says, "it is our ignorance of natural things that causes all our admiration, and chiefly excites our passions"' (Humboldt, 1997, p. 40).

10 In a letter to his Genèvan friend Marc-Auguste Pictet in 1805, he writes: 'My attention was directed to phenomena of every description, and above all it appeals to the imagination. The world likes to *see*, and I there exhibit a microcosm in a leaf'. Here science and emotion are combined to enable the reader to 'see' the wonders of the world (cited by McCrory, 2010, p. 108).

difficulties facing Darwin is to prove his theory by providing evidence of intermediate forms but he openly admits that it is difficult to do this due to the very nature of development; that is, an intermediate form will move to another intermediate form until a new species is formed, and, as intermediate forms are relatively short-lived, though the whole process is very slow, it will be difficult to find evidence of the whole chain of forms:

> It is just possible by my theory, that one of two living forms might have descended from the other; for instance, a horse from a tapir; and in this case direct intermediate links will have existed between them. But such a case would imply that one form had remained for a very long period unaltered, whilst its descendants had undergone a vast amount of change; and the principle of competition between organism and organism, between child and parent, will render this a very rare event; for in all cases the new and improved forms of life will tend to supplant the old and unimproved forms. (Darwin, 1985, p. 293)

In the same way as nothing is fixed in the formation of species, so too with inorganic Nature. As the levels of land and sea can be shown to have oscillated over millions of years, and as generations of species must have existed over a similar length of time, the collections in geological museums inadequately represent natural history:

> During oscillations of level, which we know this area has undergone, the surface may have existed for millions of years as land, and thus have escaped the action of the sea: when deeply submerged for perhaps equally long periods, it would, likewise, have escaped the action of the coast-waves. So that in all probability a far longer period than 300 million years has elapsed since the latter part of the secondary period [...]. What an infinite number of generations, which the mind cannot grasp, must have succeeded each other in the long roll of years![11] Now turn to our richest geological museums, and what a paltry display we behold! (Darwin, 1985, p. 297)

---

11 Considering that this was written over 150 years ago without the technology at our disposal today to date the age of the earth, its rocks and fossils, and measure the movement of tectonic plates, it is truly remarkable that Darwin was able to describe this so accurately. But this is not merely a description of the age of so many rocks; in stating their age it is also in parallel stating the age of the existence of species that have been found as fossils in these rocks, and in their transmutations reveal that their existence stretches back generation after generation to almost the beginning of time itself.

In what way can Darwin's collection of geological evidence towards his theory of natural selection be regarded as Romantic or to have been influenced by the Romantics? Although the evidence is critical to his theory, it is more the mental framework in which it is collected that makes it Romantic and this is where the influence of Humboldt can be seen (as already demonstrated). When Humboldt gathered his data on his South American expedition, he did so objectively through his scientific collection of data and scientific measurements, and, subjectively, through his own personal empirical experience. By 'experience' is not meant merely the perception of sense data but the overall interpretation of both the scientific data and the impressions that the perceptions made on him and how these related to other experiences he had had or others had had. Humboldt not only measured the height of volcanoes and the temperature of their rocks, not only analysed their rock structure, but literally leaned over their craters breathing in their dangerous gases. In climbing high mountains, he not only measured their height with altimeters but also measured his altitude according to the amount of bleeding his mouth was subjected to. He truly experienced Nature and science *personally* (or subjectively) as the title of his work *Personal Narrative* implies. But he did not stop with personal data collection – he related it to data collected in other regions and made comparisons. In this way he combined the subjective with the objective, the artistic narrator with the scientist. In this respect he could be seen as a 'Romantic empiricist'. So too with Darwin. He had learnt from Humboldt how to combine the personal with the objective. He was able to rise up above the personal collection of data and relate it to other data to create a holistic view of the development of the Earth and its species. But importantly he did not allow the scientific analysis to overshadow his personal experience of the perceptions or impressions. This subjective analysis comes across through his imagination of what could be missing in the evidence before him and why it could be missing. This enables him to use this as an argument *for* natural selection rather than against it; on their own, the small number of fossils in museums might be used as an argument against natural selection. The action of the oscillations affects the preservation of the evidence of transitional species as well as the destruction and extinction of life. Such geological records of these periods are therefore bound to be imperfect:

> These periods of subsistence would be separated from each other by enormous intervals, during which the area would be either stationary or rising; whilst rising, each fossiliferous formation would be destroyed, almost as soon as accumulated, by the incessant coast-action, as we now see on the shores of South America. During the periods of subsistence there would probably be much extinction of life; during the periods of elevation, there would be much variation, but the geological record would then be least perfect. (Darwin, 1985, p. 308)

From these oscillations, Darwin concludes that during periods of subsidence the number of inhabitants would decrease and the fossil remains would increase due to the preserving properties of the deposits. On the other hand, during periods of elevation of the land shallow parts of the sea would increase making it favourable for the increase in new species but not favourable for their preservation as fossils, thereby leaving a blank geological record (Darwin, 1985, pp. 301–2). Gaps in the geological record could also give the false impression of the formation of new species when species have been forced to migrate due to oscillations in land and sea, and, then, after a long period of time, migrate back to their original habitats (Darwin, 1985, pp. 308–9).

Through Darwin's reflections on Nature, from his own observations and from evidence provided by other naturalists, he is able to break with the tradition of viewing Nature as static. This is central to his argument. He is able to take this bold step by asking the question whether we have 'any right to assume that things have thus remained from eternity' and poses the possibility 'that continents may have existed where oceans are now spread out; and clear and open oceans may have existed where our continents now stand' (Darwin, 1985, p. 315). At the same time he notes the imperfect condition of the geological record which necessitates his intellectual reflection on what the gaps in the records might tell us (Darwin, 1985, p. 316). From his knowledge of geology, Darwin is also able to explain why, apart from migration, palaeontological collections are so poor. This can be explained by the softness of organisms and the action of rainwater that can dissolve the beds in which the fossils lie (Darwin, 1985, p. 298).

This Romantic concept of mental 'reflection' contributing to Man's understanding of the world is more than just an idea; it is a mental enabler helping to bring Man closer to Nature, removing barriers and prejudices that prevent Man from seeing Nature in the raw for what it is. 'Reflection' is not just thinking about the experience of the moment but putting it in the context of all other experiences in time and place and relating them to each other. At an intellectual level this rational empiricism is Darwin's form of Romantic biology (and as such can also be called Romantic empiricism as previously suggested). As mentioned earlier with reference to Humboldt, 'reflection' is contemplating isolated data, making comparisons between data collected in different regions at different times by the same or different naturalists. But the 'reflection' is not just a reflection on data, it is also a 'reflection' on the subjective sensations and impressions received from Nature. The objective and subjective are combined to create the Romantic empiricist (or the Romantic rational empiricist as the mental faculties collect, compare and analyse data coming from both objective and subjective sources). For Darwin, reflection enables him to grasp the concepts of change, time and variations, making him move away from a 'plan of creation' that only creates a circular argument:

> The chief cause of our natural unwillingness to admit that one species has given birth to other and distinct species, is that we are always slow in admitting any great change of which we do not see the intermediate steps [...]. The mind cannot possibly grasp the full meaning of the term of a hundred million years; it cannot add up and perceive the full effects of many slight variations, accumulated during an almost infinite number of generations [...]. It is so easy to hide our ignorance under such expressions as the 'plan of creation,' 'unity of design,' &c., and to think that we can give an explanation when we only restate a fact. (Darwin, 1985, pp. 452–3)

Darwin's 'reflective' powers can be seen to be part of a cyclical process. Darwin's first-hand experience of Nature and his reading of geologists of the time, such as Charles Lyell's *Principles of Geology* (1830–1833)[12] (Lyell, 1997) which he took with him on the voyage of the *Beagle*,

---

12   'When [I was] starting on the voyage of the *Beagle*, the sagacious Henslow [...] advised me to get and study the first volume of the *Principles*, [...] but on no account to accept the views therein advocated. How differently would anyone now speak of the *Principles!* I am proud to remember that the first place, namely, St Jago, in the Cape de Verde Archipelago, in which I geologised, convinced me of the infinite superiority of Lyell's views over those advocated in any other work known to me' (Darwin, 1995, pp. 33–4).

enabled him to put his experiences within the context of oscillations of land and the preservation of fossils, etc. This 'reflection' then created another form of 'reflective' glasses to examine Nature with, but these 'reflections' were never fixed as they were continually being adjusted according to further experiences of raw data gained from Nature. In short he was a true modern scientist testing out his theories in the field, but, at the same time, a Romantic natural philosopher, not letting his objective science become divorced from his subjective personal experiences of Nature obtained first hand (or second hand via the experiences of other naturalists and geologists). This is illustrated when Darwin refers to Lyell and Dawson's discovery of carboniferous beds with recurring deposits of the same fossil remains, making him 'reflect' and conclude that the species concerned could not have lived in the same place for the whole period and that therefore they must have appeared and reappeared many times (Darwin, 1985, p. 304). Like Humboldt before him, he is sifting, collating and comparing data from his own experiences and the experiences of other naturalists. As a natural *historian* he is developing the *narrative* of creation from the beginning of species by theorizing on how they developed according to the probability that the evidence provides. This can be seen in the tentative use of his words such as 'probability', 'perhaps', 'if such species were to', and 'a section would probably not include'. The raw data is the 'objective science', the imaginings are the 'subjective science', yet together they create a Romantic empirical science. As explained before, Nature can be seen as 'Mind' if it is understood to be a reflection of what the perceiver sees through his sense experiences, *and* a reflection of Man's understanding through his rational powers of theorizing about the world as science; Nature can therefore be seen to be Romantically rational *and* Romantically empirical.

By relating these 'complex contingencies' through his 'reflective' thinking, Darwin is able to formulate his theory of natural selection. Although these are abstract mental constructs, they refer to laws in Nature enabling species to be mutable. The mental relationships of the constructs making up Darwin's theory mirror a corresponding set of laws actually existing in Nature driving changes in species to come about. In this sense Nature is organic and therefore Romantic: the abstract and the actual combine to form the whole. This interrelatedness of the parts of Nature is also reflected in Darwin's 'reflective thinking'. This conceptual framework of his thinking that makes up his theory of natural selection is also connected up to form a 'web of affinities'; for example, the concepts of time and place, slowness of development, appearance of new forms and disappearance or extinction of old forms:

> The theory of natural selection is grounded on the belief that each new variety, and ultimately each new species, is produced and maintained by having some advantage over those with which it comes into competition; and the consequent extinction of less-favoured forms almost inevitably follows. It is the same with our domestic productions: when a new and slightly improved variety has been raised, it at first supplants the less improved varieties in the same neighbourhood [...]. Thus the appearance of new forms and the disappearance of old forms, both natural and artificial, are bound together. (Darwin, 1985, p. 323)

Darwin argues that his theory of slow development shows that species change over time according to advantage. They do not *necessarily* diverge from those that came before as divergence only depends on what advantages can be gained for the species. This depends upon the conditions of the habitat and its inhabitants. Therefore some are more modified than others and exist longer than others (Darwin, 1985, p. 322). On his travels to the Galapagos Archipelago, Darwin noted that not all migrant species and aboriginal species modified themselves if no advantage was required. In other words, certain aboriginal and migrant species could live side by side without being in direct competition with each other (Darwin, 1985, p. 389).

Another example of Darwin's 'reflective' thinking can be seen in his realization whilst on the Galapagos that migrating species from distinct genera can adapt themselves to their habitat in the same way as the aboriginal species. Some migrants may give the appearance of being related to the aboriginals, but in fact are species from different genera. This just demonstrates that both species have adapted themselves to their habitat in a way which is the most advantageous to them, and that both have found the same advantage without being related to each other.[13] This may seem obvious but actually demonstrates Darwin's 'reflective' thinking that goes beyond specific sense data. Someone comparing two similar animals or plants may assume they come from the same progenitors, but this is not

---

13   This is now called 'convergent evolution'.

necessarily the case (Darwin, 1985, p. 389). Darwin's reading of Lyell and his own experience of different geographical regions enabled him to make comparisons between species that were the same and species that had modified themselves over time. From this data he was able to determine that similarities were due to 'bonds of inheritance' and dissimilarities were due to migration with modifications as well as barriers preventing migration such as mountains and large rivers. Seeing species of the same genus in different areas convinced Darwin that they must have migrated.

Through his 'reflective' thinking, Darwin comes back to Nature to carry out his own experiments to test out his theories of migration. He carries out all kinds of experiments on seeds to see if they can be transported by driftwood, dead bird crops, bird excrement, beaks, feet and icebergs, and whether they can maintain their powers of germination (having been immersed in sea water for long periods of time). He does similar experiments on snails (Darwin, 1985, pp. 354–8). On reflecting on the similarities of species on mountain tops in the United States and Europe, Darwin is able to theorize on the effect of the ice age on species distribution and concludes that plant and animal species would move up and down the mountains according to the temperature changes as well as mingle with the ancient species thus creating modifications (Darwin, 1985, pp. 360–2).

These important comparisons (or forms of 'reflective' thinking) hark back to Humboldt's establishment of plant geography in which he not only compares the vegetation at different heights on one mountain, but compares the vegetation at the same heights on different mountains, noting their similarities and differences. As he was climbing the volcano of Tenerife he realized that the principles of vertical geography were the same as the principles of horizontal geography, and this developed his 'habit of viewing the globe as a great whole' (Sachs, 2007, p. 52). His sketches and commissioned plates show the geographical distribution of plants in detail, including such factors as sea level, temperature and the exposure to other plants.[14] The plates are not just a snapshot of physical data, but a plant geography experienced personally, as his own body experienced the changes in altitude (see Figures 1 and 2).

---

14 First published in Paris in 1825, and reproduced in Lack, 2009, pp. 47–8: plate 28 Geography of the Plants near the Equator, 1803, profile of the Andes from west to

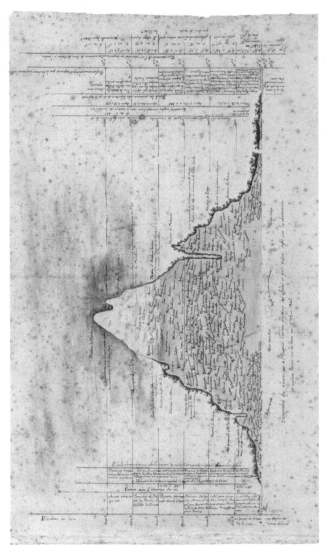

Figure 1: A.v. Humboldt, *Geography of the Plants near the Equator*, 1803. © Museo Nacional de Colombia/Oscar Monsalve Pino. Permission to reproduce the photo kindly granted by Museo Nacional de Colombia and Oscar Monsalve Pino. Colección Museo Nacional de Colombia, reg. 1204.
Alexander von Humboldt (1769/1859). Geografía de las plantas cerca del Ecuador.
Tabla física de los Andes y países vecinos, levantada sobre las observaciones y medidas tomadas en los lugares en 1799–1803, 1803. Acuarela (Acuarela y tinta Papel) 38.7 × 50.3 cm.

Darwin's Romantic Theory of Mind

Figure 2: L. A. Schönberger and P. J. F. Turpin after A. v. Humboldt and A. Bonpland, *Géographie des plantes équinoxiales*, 1807. Permission to reproduce image from book kindly granted by Peter H. Raven Library/Missouri Botanical Garden and the Biodiversity Heritage Library <http://www.biodiversitylibrary.org>.

Essai sur la geographie des plantes: accompagne d'un tableau physique des regions equinoxiales, fonde sur des mesures executes, depuis le dixieme degree de latitude boreale jusqu'au dixieme degree de latitude austral, pendant les annees 1799, 1800, 1801, 1802 et 1803/par Al. de Humboldt et A. Bonpland; redigee par Al. de Humboldt. A Paris, Chez Levrault, Schoell et compagnie, libraires, XIII-1805. <http://www.biodiversitylibrary.org/bibliography/9309>.
Item: <http://www.biodiversitylibrary.org/item/37872>, page 156.

There is a feeling that as he experiences these changes in Nature more deeply, he becomes more subjected to Nature rather than the explorer imposing himself on Nature. In this position of immersion he is able to experience the whole, yet, at the same time, as a scientist taking measurements, he is also able to describe it objectively. Humboldt's plates can be viewed both as works of beauty and art displaying the wholeness and unity of the mountain ranges in relation to other ranges and their vegetation, and as works of scientific research displaying the detailed data of their height and plant names (Botting, 1973, pp. 207–8; McCrory, 2010, p. 123). So too with Darwin, but for him his imagination goes beyond the personal and objective experiences of the plant geography of mountains (whether this be in the form of art or science); for Darwin it is the uncovering of the secret of Nature that Humboldt only caught a surface glimpse of. For Darwin it was not the overview of Nature from the height of a mountain top (or comparatively of the mountain ranges through sketches) as with Humboldt; it was the secret of the inner force of Nature through its inner laws in the form of natural selection. For Darwin, the more he scratches away at the surface of Nature, the more its secrets are revealed.

From 'tide marks' left on mountain tops (another 'rastro' telling a story), and with inspiration from Lyell, Darwin is able to understand how species can get stranded from rising waters and how this can combine with natural selection to create great cycles of change (Darwin, 1985, pp. 372–3):

> As the tide leaves its drift in horizontal lines, though rising higher on the shores where the tide rises highest, so have the living waters left their living drift on our mountain-summits, in a line gently rising from the arctic lowlands to a great height under the equator. The various beings thus left stranded may be compared with savage races of man, driven up and surviving in the mountain-fastnesses of almost every land, which serve as a record, full of interest to us, of the former inhabitants of the surrounding lowlands. (Darwin, 1985, pp. 372–3)

---

east, ink and watercolour on paper; plate 29 Geógraphie des plantes équinoxiales, 1807, profile of the Andes, coloured engraving, Paris; plate 30 Geographiæ plantarum lineamenta, 1815; plate 31 Vegetation profile on Chimborazo, 1825, coloured stipple engraving.

These 'tide marks' (or 'rastros') narrate their histories, both physically and metaphorically. They express Darwin's deep 'reflective' insights into Nature and it is clear from the above that Darwin is not merely restating fact. By relating the stranded species to stranded 'savage races of man' he brings home the feeling of isolation and survival that is part of the struggle of existence and the concept of natural selection. In referring to the 'savage races of man' being 'driven' up the mountains there is an extra emphasis on the concept of struggle as Darwin witnessed both natives suffering hardships due to their difficult habitats and slaves suffering brutal treatment from their masters. In both cases Man could be seen to be driven away from his own habitat and into isolation. Well after Darwin, remains of isolated cavemen have been discovered supporting this very image and today, during extended periods of drought in countries such as Sudan, thousands of families die during mass migrations in search of food and water. Whereas Humboldt mapped a plant geography, Darwin mapped a genealogical plant *and* palaeontological geography revealing the history of Nature through its development in time and place. Darwin's famous 'tree diagram' showing the development of species, is in a sense a 3D version of Humboldt's plant geography plates. Darwin goes one step further by digging deeper to show their historical relationships rather than just the way they are now and extending this to cover all life. But the experience is just as Romantic, if not even more so, in relating all the mental constructs making up this image of Nature.

Linked to Darwin's 'reflective' thinking is the mind's use of archetype. Its use can be argued to be Romantic because the archetypal forms and structures *imaginatively* capture the organic whole of the essence of Nature (in the same way as the union of science and poetry capture this organic whole). Darwin's use of the concept of archetype will now be examined in greater detail in relation to Goethe's archetypes (briefly referred to earlier) in order to shed more light on Darwin's 'reflective' and imaginative thinking. Richards (2002) argues that Romantic thinkers such as Goethe and Schelling believed that Nature was organically composed of archetypes that developed towards their ideal forms (Richards, 2002, p10). They argued against Kant's *intellectus archetypus*[15] in which the Divine mind was the

---

15 Outlined in Kant's introduction to his third *Critique*, the *Critique of Judgement* (cited by Richards, 2002, p. 68). See also Kant, 1914, unnumbered page, IX, 39.

sum total of archetypes (or ideas) and against an objective 'noumenal' thing-in-itself world of objects existing separately from the human world of experience. Goethe developed the notion of *Urtypus* (or formative force or *Bildungstrieb*) which was the fundamental archetype that included the *Urpflanze*, or plant archetype, and the vertebrae archetype. The *Urpflanze* was the archetype of all plants. The idea of the 'transcendental leaf' existed in all seeds. As the plant developed from the stem so would the leaf develop into flowers and then into seeds again. This involved the forces of repulsion and attraction as the leaf developed from one thing into another. A development in one resulted in a reduction in the other and Goethe called this the 'law of compensation'.

The idea of the *Urpflanze* came to Goethe while walking in the Public Gardens of Palermo, Sicily, on the 17 April 1787 during his *Italian Journey* (1786–1788). Seeing plants in their natural environment enabled him to see their natural forms:

> Here where, instead of being grown in pots or under glass as they are with us, plants are allowed to grow freely in the open fresh air and fulfil their natural destiny, they become more intelligible. Seeing such a variety of new and renewed forms, my old fancy suddenly came back to mind: Among this multitude might I not discover the Primal Plant? There certainly must be one. Otherwise, how could I recognize that this or that form *was* a plant if all were not built upon the same basic model? (Goethe, 1970, pp. 258–9)

During his second Roman visit on 31st July 1787, he refers back to his visit to the gardens saying that the idea of the '*leaf*' as the Primal Plant came to him 'in a flash':

> While walking in the Public Gardens of Palermo, it came to me in a flash that in the organ of the plant which we are accustomed to call the *leaf* lies the true Proteus who can hide or reveal himself in all vegetal forms. From first to last, the plant is nothing but leaf, which is so inseparable from the future germ that one cannot think of one without the other. (Goethe, 1970, p. 366)

This insight made him realize that the archetype could not be seen empirically but could only be understood intellectually. He also realized that there was a commonality amongst the form of all plants that made them recognizable as plants and that all the parts of a single plant had a similar structure (Miller, 2009, p. xviii). Miller agrees with Richards that Goethe echoes Spinoza's holistic view of Nature, citing Goethe's view that we can only understand the inner laws of Nature if we employ the eyes of the mind as well as the eyes of the body, that is, by employing 'sensory and intuitive perception', ensuring a balance between the 'human spirit' and 'the informing spirit of nature' (Miller, 2009, p. xviii). Goethe, like Spinoza, 'coupled rigorous empiricism with precise imagination to see particular natural phenomena as concrete symbols of the universal principles, organizing ideas, or inner laws of nature' (Miller, 2009, p. xviii). This is key to the understanding of the inner archetypal nature of things: starting with a sensory perception of the world (empiricism) but then 'seeing' the particular instances of objects as 'symbols of the universal principles' (imagination); that is, insight into the secret laws of Nature. For Humboldt his imagination took the form of aesthetic understanding; for Darwin it took the form of the law of natural selection. Both employed 'the eyes of the body' and 'the eyes of the mind'. In this context the physical and the mental aspects of Nature become a unified whole, striving for perfection on the one hand, yet through human imagination on the other hand bringing out the forms of the structures, almost an act of turning objects inside out revealing their hidden natures. For Goethe this revelation of their hidden natures can be seen in his understanding of the development or metamorphosis of plants; he saw this as depending on two processes, *'intensification'* (the striving for perfection) and *'polarity'* (the state of constant attraction and repulsion like electricity and magnetism) – seeds develop into plants and then form seeds again. But this process is dialectical in that the process depends on both the inner law of Nature as well as the external law of the environment[16] (as with Humboldt's 'plant geography'). The inner potential or formative force of organic beings (*Proteus in potential*) is realized or actualized through organic forms such as leaves, petals, backbones, etc. (*Proteus actus*).

---

16 Goethe gives examples of leaves adapting to different environmental conditions: 'The leaves of underwater plants likewise show a coarser structure than those of plants exposed to the open air; in fact, a plant growing in low-lying, damp spots will even develop smoother and less refined leaves than it will when transplanted to higher areas, where it will produce rough, hairy, more finely detailed leaves' (Goethe, 2009, p. 19).

But the actual structures adapt to the environmental conditions (*Proteus actus adaptatus*).[17] This is the Romantic notion of wholeness and unity in Nature that involves 'the interdependence of organism and environment, as well as organism and organism' (Miller, 2009, pp. xx–xxi). This unity of Nature, according to Goethe, can also be seen as a unity of understanding in Man expressed through the union of science and poetry (also endorsed by Humboldt).[18] In addition to his essay, Goethe also wrote a poem on 'The Metamorphosis of Plants' in which this union of science and poetry is clearly expressed through his description of the secret laws of Nature, the power of development in the seed with its pre-formed forms:

> Like unto each the form, yet none alike;
> And so the choir hints a secret law [...]
> Gaze on them as they grow, see how the plant
> Burgeons by stages into flower and fruit,
> Bursts from the seed so soon as fertile earth [...]
> Asleep within the seed the power lies,
> Foreshadowed pattern, folded in the shell,
> Root, leaf, and germ, pale and half-formed.[19]

This idea or archetype was not a thing that could be seen empirically in particular objects, but could only be seen by the inward eye (Richards, 2002, p. 452). This was the same for the animal archetypes but, unlike the plants, reversal in development was not possible. For Goethe, form is about becoming and transformation ('Asleep within the seed the power lies'). For him this is the key to all Nature, seen and not seen ('Foreshadowed pattern, folded in the shell'). Inorganic forms such as rocks are developed passively whereas the organic forms of Nature are able to reproduce themselves (Richards, 2002, pp. 453–4). Although Man can experience the forces of

---

17   Miller, 2009, p. xxi. See also Bortoft, 2010, pp. 275–89.
18   'The great naturalist Alexander von Humboldt dedicated an 1806 book to Goethe with an illustration featuring *The Metamorphosis of Plants* and imagery suggesting, true to Humboldt's Romantic sympathies, that poetry as well as science can succeed in uncovering the secrets of nature' (Miller, 2009, p. xxiv).
19   Extract from Goethe, 1949, cited in Goethe, 2009, p. 1.

Nature empirically through experience, Goethe believed that the whole experience of life could only be experienced through the non-physical eye, and this applied both to the experience of Nature and the experience of the self. The becoming of form (for example the shape and size of an animal's neck or tail) was the result of the interaction between the archetype or *Bildungstrieb* and the external forces in the environment. Through his poem 'The Metamorphosis of Plants' and his essay *The Metamorphosis of Plants* (1790) (Goethe, 2009), Goethe believes he can uncover the archetypes that express this force of Nature (Richards, pp. 466–7). Whereas for Goethe the force of Nature was the sum total of eternal archetypes seeking development towards their perfection, Darwin's force of Nature was natural selection. Yet there is a similarity in that for both Darwin and Goethe, modifications are the result of interactions between forms and their environment. For both, despite these modifications, the essential internal forms or archetypes remain the same. As with Darwin's 'incipient species' in the development of species, Goethe identifies intermediate forms in the metamorphosis of plants in which change cannot be achieved in one step and must be produced through 'intermediate agents' as 'gradual transitions':

> However rapid the transition from corolla to stamens in many plants, we nonetheless find that nature cannot always achieve this in a single step. Instead, it produces intermediate agents that sometimes resemble the one part in form and purpose, and sometimes the other. Although they take on quite different forms, almost all may be subsumed under one concept: they are gradual transitions from the petals to the stamens. (Goethe, 2009, pp. 44–5)

For Goethe the key to understanding Nature is becoming and transformation. Darwin was familiar with Goethe as he is referred to in the *Origin*, for example when discussing the 'law of compensation':

> The elder Geoffroy and Goethe propounded, at about the same period, their law of compensation or balancement of growth; or, as Goethe expressed it, 'in order to spend on one side, nature is forced to economize on the other side'. (Darwin, 1985, p. 185)

Becoming and transformation are also regarded by Darwin as a key to understanding Nature and this is evident when Darwin talks about flowers being 'metamorphosed leaves', an obvious reference to Goethe's 'leaf' archetype outlined in his essay *The Metamorphosis of Plants* (and his poem of the same title):

> It is familiar to almost everyone, that in a flower the relative position of the sepals, petals, stamens, and pistils, as well as their intimate structure, are intelligible in the view that they consist of metamorphosed leaves, arranged in a spire. (Darwin, 1985, p. 417)

This reference to the metamorphosis of leaves is almost exactly the same as Goethe's, showing that there is at least some influence from Goethe in terms of the archetypes of plants and animals.

For Darwin, the evidence of gradual development reveals the laws of natural selection. Gradual development can be seen through the endurance of species in time and space. Modifications over time hide the essential blood ties between species with their essential hidden forms remaining unchanged. All are linked together through space and time, and their endurance can be seen in deposits in rock strata (Darwin, 1985, p. 395).

This endurance of species' blood ties is due to the essential hidden forms remaining unchanged and is summed up by Goethe as 'their inner essence':

> The different plant parts with their apparent variety of forms are nonetheless identical in their inner essence. (Goethe, 2009, p. 54)

The 'inner essence' as archetype is the same whether the plant part is the leaf, the flower, the fruit (or the seed), and is therefore the same throughout the reproductive cycle:

> Whether the plant grows vegetatively, or flowers and bears fruit, the same organs fulfil nature's laws throughout, although with different functions and often under different guises. The organ that expanded on the stem as a leaf, assuming a variety of forms, is the same organ that now contracts in the calyx, expands again in the petal, contracts in the reproductive apparatus, only to expand finally as the fruit. (Goethe, 2009, p. 100)

Central to my argument in demonstrating a link between Goethe and Darwin is the use of the Romantic concept of the archetype. But the link goes beyond the mere use of the archetype. It can also be argued that the *method* of its use is also Romantic; that is, how the mind through its overview of forms and structures is able to *imaginatively* capture the organic whole of the essence of Nature. Goethe believed that science, poetry and art could jointly uncover the secret essence of Nature. He had already written an essay and a poem on 'The Metamorphosis of Plants' and was hoping one day to illustrate his essay but never got round to it (Miller, 2009, p. xxviii). He believed that illustrators, as well as scientists and poets, could also capture the essence through light, shadow and perspective (Miller, 2009, p. xxviii). This is why Miller added his photographic illustrations to his edited edition of Goethe's *Metamorphosis of Plants* (Goethe, 2009).[20] But the illustrations do much more than just illustrate the plant. They show how the lower leaves of a plant metamorphose into the upper leaves and by the observer moving their eye up and down the plant they are able to understand the archetype or essence of *all* plants. Goethe calls the method for finding out how the diversity of physical forms emerges from their underlying unity, the 'Genetic Method'. This does not refer to the science of genes as we know it today but to 'the origin or genesis of things' (Miller, 2009, p. 105). My argument is that Darwin also uses this method when identifying common forms by winding back (genealogically) the developmental process through time and space and then forwards (teleologically), again and again, like rewinding and forward-winding a film in order to understand its concepts better. Through seeking the origin of the creation process, both Goethe and Darwin are seeking out the steps that make the diversity of creation possible. This is summed up by Goethe as follows:

> If I look at the created object, inquire into its creation, and follow this process back as far as I can, I will find a series of steps. Since these are not actually seen together before me, I must visualize them in my memory so that they form a certain ideal whole. At first I will tend to think in terms of steps, but nature leaves no gaps, and thus, in the end, I will have to see this progression of uninterrupted activity as a

---

20 'It is my hope that the present illustrated edition, while not the full sequel that Goethe envisioned, will nevertheless aid the metamorphosis of that tree of knowledge toward diverse and wide-spreading foliage and particularly deep roots', Miller, 2009, p. xxix).

whole. I can do so by dissolving the particular without destroying the impression itself. (from Goethe, 1988, p. 75, cited by Miller, 2009, p. 105)

This method enables the observer to move backwards and forwards between the relatively fixed forms and the formative processes, between the underlying unifying archetypal essences and the particular individuals. As previously discussed, this is the movement between producer and product, or, as Spinoza would put it, *Natura naturans* (*'Nature naturing'*) and *Natura naturata* (*'Nature natured'*). It is the imaginative ability of the 'Genetic Method' that enables the observer to understand the essence of Nature through its laws and through its unified form of creator and creation. For Goethe this imagination (of creator and creation) is expressed through the line in his poem 'gaze on them as they grow' (Goethe, 2009, p. 1) and, for Darwin, it is expressed in his description of the 'entangled bank' that sums up his theory of natural selection. Miller explains this imaginative process of the 'Genetic Method' through reference to his illustrative photos showing the metamorphic process. According to Goethe the 'Genetic Method' involves understanding the sequence of changing forms as well as the 'exact sensory imagination'. In his 'Image 55: Leaf sequence in *Sidalcea malviflora*', Miller explains the sequence of changing form by showing four leaves developing upwards from a rounded shape to a less rounded shape with incisions (see Figure 3). By looking at all the leaves as a whole we can see the 'sameness in the midst of the differences':

> The leaves become larger and less rounded, the incisions grow into definite divisions, but the original plan is still evident in the pattern of the veins. Thus there is sameness in the midst of the differences. (Miller, 2009, p. 106)

The 'exact sensory imagination' is achieved by reviewing the sequences, then internalizing them as visual images, and then through the imagination going backwards and forwards between them. The mind is then able to transform one into the other, from formative to finished form, and, back again, from finished form to formative form. This is where the process of the metamorphosis of the plant becomes alive in the imagination of the mind, where the Mind and Nature also become unified. As expressed by Miller:

> What was successive in one's empirical experience then becomes simultaneous in the intuitively perceived idea – *Proteus in potentia*. Instead of an onlooking subject

Figure 3: Leaf Sequence in Sidalcea Malviflora. From Miller, 2009, Image 55, p. 107. © Gordon L. Miller and The MIT Press, Cambridge, MA, USA. Permission to reproduce the photo kindly granted by Gordon L. Miller and The MIT Press.

> knowing an alien object, this is knowledge through participation, or even identification, of observer and observed – knowing things from the inside. (Miller, 2009, p. 111)

I would like to argue, therefore, that Darwin's classification system of Nature is not just a naming system as created by Linnaeus but a system that reveals the process and community of descent through the genealogical 'tree of life' using the 'Genetic Method'. The 'tree of life', I would argue, is a 'Genetic Method' (a form of 'reflective' imaginative thinking of the highest level) as it portrays the relationships of the morphological process going both forwards and backwards in time. The arrangements of the species in the tree reveal the degree of modifications over time and are arranged accordingly as are the intermediate groups. This is seen by moving backwards and forwards between the particular and the underlying forms. The changes over time highlight the dominant groups which make up the largest orders and these reflect those that have been successful or those that have failed and become extinct. Referring to the diagram in Chapter 4 of the *Origin*, pp. 160–1, Darwin expresses this as follows:

> All the modified descendants from A will have inherited something in common from their common parent, as will all the descendants from I [...]. If, however, we chose to suppose that any of the descendants of A or of I have been so much modified as to have more or less completely lost traces of their parentage, in this case, their places in a natural classification will have been more or less completely lost [...]. All the descendants of the genus F, along its whole line of descent, are supposed to have been but little modified, and they yet form a single genus. But this genus, though much isolated, will still occupy its proper intermediate position; for F originally was intermediate in character between A and I [...]. (Darwin, 1985, p. 405)

This genealogical community of descent is both relational and reflective as previously outlined. It is relational in that all the branches making up the tree are related through space and time, and this can be seen not only in fossil deposits as a trace of their previous existence but also in their similarities to existing and extinct species (that is, similarities between fossil remains). It is also relational in terms of the traces of past migrations and habitats and co-existing species in those habitats that make up the history or narrative of their past existences. Understanding the archetypes that form the species and understanding the relationships between them requires reflection and this enables the biologist or naturalist to understand the whole. This 'reflection' is the 'Genetic Method'; it is about understanding the relationships going both forwards and backwards. In Kantian terms this reflection is looking at the structures in Nature *as if* the archetype or archetypes created them, *as if* Nature or Nature's God had created them according to some teleological ideal. This form of reflection is pragmatic in that it is used to help make sense of the world but without assuming that the archetypes have actually caused the structures as this cannot be empirically proved (Richards, 2002, p. 489). One of Darwin's insights equivalent to Goethe's discovery of the 'leaf' and 'skull' archetypes, was his discovery that species that were distinct from each other in adulthood were less distinct in their junior or embryonic forms. This insight came to him when examining barnacle larvae, finding that they were similar to crabs and lobsters at the same stage of development.[21] This insight is the realization that all creatures are built from common structures or archetypes. Understanding this similarity intellectually through the 'Genetic Method' enables him to understand Man's genealogical heritage and his humble beginnings. As expressed by Steve Jones, all creatures are built using a similar modular structure which is 'shuffled around' as 'growth proceeds':

> We do not often think of ourselves as segmented creatures, but the vertebrate body is, like that of a barnacle or a lobster, also based on a series of distinct units, arranged from front to back. The human head, thorax and abdomen are obvious enough but our muscles, or our brain-case, show little sign of order. A glance at the embryo, however, reveals that men and women, like their submarine relatives, are constructed from a series of modules, neatly arranged in early life but shuffled around and modified as growth proceeds. (Jones, 2009, p. 200)

---

21   Darwin was excited by the discovery of this similarity describing one barnacle lava 'with six pairs of beautifully constructed natatory legs, a pair of magnificent compound eyes, and extremely complex antennæ'. As highlighted by Jones, 'he knew that he had hit upon a crucial piece of evidence for evolution (although his children laughed because the sentence read like a newspaper advertisement by a cirripedes manufacturer)' (Jones, 2009, p. 199).

For Darwin the archetype is seen as an unchanging form carried forwards from an ancient progenitor. The essential inner hidden form is unchanging yet the forms seen in Nature modify themselves according to competing species and environmental conditions. This seems like a contradiction. How can a form remain unchanged yet change? When talking about unchanging forms he is referring to the unchanging nature of an archetype such as a mouth. The essential underlying function of a mouth is the same whether it is of a snake, an insect, a fish or a bird. Although the structures of the mouths may vary according to the type of species, due to the things they eat and the environment in which they live, the underlying (archetypal) function is the same. Although biologically the structures of insect and mammal mouths are different, archetypally in terms of function their underlying structure can be regarded as the same. This also applies to bones and eyes. Their essential archetypes remain the same but their particular forms will also vary:

> The bones of a limb might be shortened and widened to any extent, and become gradually enveloped in thick membrane, so as to serve as a fin; or a webbed foot might have all its bones, or certain bones, lengthened to any extent, and the membrane connecting them increased to any extent, so as to serve as a wing: yet in all this great amount of modification there will be no tendency to alter the framework of bones or the relative connexion of the several parts. If we suppose that the ancient progenitor, the archetype as it may be called, of all mammals, had its limbs constructed on the existing general pattern, for whatever purpose they served, we can at once perceive the plain signification of the homologous construction of the limbs throughout the whole class. So with the mouths of insects, we have only to suppose that their common progenitor had an upper lip, mandibles, and two pair of maxillae, these parts being perhaps very simple in form; and then natural selection will account for the infinite diversity in structure and function of the mouths of insects. (Darwin, 1985, pp. 416–17)

However, Darwin does admit that it is possible that the patterns of an organ may be modified to such an extent that its original form may be lost implying that it may be difficult, through these losses, to trace their origins to their archetypes and their primordial progenitors (Darwin, 1985, pp. 416–17).

But in the main it should be possible to trace the origins as any modifications of organs or species will have taken place very slowly over a long period of time thereby making it possible to identify these modifications in the intermediate forms, or incipient species, before becoming new species. These changes would be represented by slight successive steps leaving a thread of resemblance through each stage which would be passed on through inheritance. Darwin, like Goethe before him, stressed that both the vertebrae and leaf archetypes had homologous structures that were maintained throughout their modifications of forms for the purpose of different functions. As discussed earlier, Goethe's moment of inspiration when coming across a dead sheep's skull was the realization that the bones were in sections built up in basically the same way as the vertebrae making up the back or neck of an animal. This meant, therefore, that the vertebrae archetype was the same wherever it was located, although the form varied according to function.

Darwin shares the same insight when comparing different parts or organs in the same individual, noting how this applies to animals and plants:

> Most physiologists believe that the bones of the skull are homologous with – that is correspond in number and in relative connexion with – the elemental parts of a certain number of vertebrae. The anterior and posterior limbs in each member of the vertebrate and articulate classes are plainly homologous. We see the same law in comparing the wonderfully complex jaws and legs in crustaceans. It is familiar to almost every one, that in a flower the relative position of the sepals, petals, stamens, and pistils, as well as their intimate structure, are intelligible in the view that they consist of metamorphosed leaves, arranged in a spire. In monstrous plants, we often get direct evidence of the possibility of one organ being transformed into another; and we can actually see in embryonic crustaceans and in many other animals, and in flowers, that organs, which when mature become extremely different, are at an early stage of growth exactly alike. (Darwin, 1985, p. 417)

The insight into the plant archetype is the same as Goethe's in which all plant forms develop from the leaf, whether they are roots or petals. So too for Darwin are 'sepals, petals, stamens and pistils' no more than metamorphosed leaves. And the reason that differences are created from common archetypes, whether in the form of organs in the individual organism such as petal or root, or between species, is simply one of natural selection:

> Why should similar bones have been created in the formation of the wing and leg of a bat, used as they are for such totally different purposes? [...]. Why should the sepals, petals, stamens, and pistils in any individual flower, though fitted for such widely

different purposes, be all constructed on the same pattern? On the theory of natural selection, we can satisfactorily answer these questions. (Darwin, 1985, pp. 417–18)

This concept of 'natural selection' is as much a mental concept as a physical one. In the vein of the 'Genetic Method', it acts as a slide rule with a set of algorithms enabling the scientist to go forwards and backwards in time and place to calculate species type historically and contemporaneously; at the same time giving insights into the variables that affect sameness or difference of function (such as altitude, habitat, cohabiting species, migration, food sources, shelter and the struggle for survival). Importantly, as Humboldt discovered, the true picture cannot be captured unless it also includes the human scientist making the observations. Man is also part of Nature, and his observations are also a mental construct that need to be included in the observations themselves.

Man's inclusive part of Nature can be seen in his need to make sense of Nature. As a reflecting animal, Man is not just living in Nature or even observing it objectively but reflecting upon it with himself a part of it. He is forever observing himself as observer whether as scientist or mere participant. For the Romantics such as Goethe and Humboldt, this is where the overlaps of science, poetry and art are important in emphasising the interdependence of these human disciplines in making sense of the laws of Nature. In Darwin, this subjectively human, or aesthetic, way of looking at Nature can be seen in the choice of words used to describe it; words such as 'struggle' or 'advantage'. Darwin talks about the organs or species developing in one way or another according to what will produce the best advantage. Darwin's theory of descent with modifications shows that the basic archetype is unchanging but varies its form according to use and advantage in a competing world of struggle. This explains why development is gradual since organs or species adapt and modify themselves according to need, a step at a time. These needs are influenced by the conditions at the time, covering everything from the habitat, the climatic conditions and other competing species; and in the case of organs, the individual's other organs and how these work together to form the whole organism. This is why the basic structure of the bones in the hand of a man are the same as that of a bat and why therefore the organs 'were alike in the early progenitor of each class':

> The framework of bones being the same in the hand of a man, wing of a bat, fin of a porpoise, and leg of the horse, – the same number of vertebrae forming the neck of the giraffe and of the elephant, – and innumerable other such facts, at once explain themselves on the theory of descent with slow and slight successive modifications. The similarity of pattern in the wing and leg of a bat, though used for such different purposes, – in the jaws and legs of a crab, – in the petals, stamens, and pistils of a flower, is likewise intelligible on the view of the gradual modification of parts or organs, which were alike in the early progenitor of each class. (Darwin, 1985, p. 451)

Despite being a scientist, Darwin cannot separate himself from the world of Nature and what it means to also be a part of it.[22] His own concepts of 'competition' and 'modification' are derived from his own experience of what it is to be human and to struggle in Nature; however much he wishes to be the objective observer, he is always going to be influenced by his human way of looking at the world; to an extent his analysis is always going to be in part anthropomorphic. Our own mental processes that are used in comparing objects to discover their differences and similarities are a necessary part of being human and are a part of our survival mechanism. We need to be able to distinguish between safe and dangerous foods, between safe and dangerous animals, and between friend and foe. But being aware of ourselves as part of Nature, being able to move from the particular to the universal by using the 'Genetic Method', makes the concept of Nature Romantic. This concept of 'gradual modification' can only be captured Romantically by mentally going backwards and forwards in time to 'see' the development through all its progressive stages and in reverse. This mental process also helps Darwin identify the gaps in development (or progress). In the same way as organs are developed through use, they are also reduced through disuse. Darwin gives examples of rudimentary teeth in a calf that do not cut through in the upper jaw in adulthood as they are not required (Darwin, 1985, p. 451).

This narrowing down of all living things to a limited number of archetypes from which they get their forms and structures also means that all living things must have come from a limited number of progenitors. Packed

---

22 Which is why Darwin prefers to refer to himself as a 'naturalist' rather than a 'scientist' as the term 'naturalist' indicates that he is not separate from Nature.

into this assertion is the assumption that all things must therefore be related through the 'web of affinities', that they developed over a very long period of time ('a hundred million years' or so), that the process is very slow as change consists of lots of steps resulting in intermediate forms before new species come into existence, and that this is not a 'miraculous act of creation':

> All the members of whole classes can be connected together by chains of affinities, and all can be classified on the same principle, in groups subordinate to groups. Fossil remains sometimes tend to fill up very wide intervals between existing orders. Organs in a rudimentary condition plainly show that an early progenitor had the organ in a fully developed state; and this in some instances necessarily implies an enormous amount of modification in the descendants. Throughout whole classes various structures are formed on the same pattern, and at an embryonic age [sic] the species closely resemble each other. Therefore I cannot doubt that the theory of descent with modification embraces all the members of the same class. I believe that animals have descended from at most only four or five progenitors, and plants from an equal or lesser number. (Darwin, 1985, p. 454)

A close examination of this text reveals a mental reconstruction of Nature built upon Darwin's theory of natural selection using the 'Genetic Method'. His use of the term 'chains of affinities' reflects the movement from past to present and present to past as the 'chains' is a connecting term that does not point in any one direction. The term 'affinities' shows similarities but these cannot be seen in isolation – both past and present need to be brought together in classifying the classes. The 'affinities' in a sense are the 'chains' creating the bridge between past and present, creating the narrative of Darwin's *Origin* (or Genesis). The fossils and the 'rudimentary' organs also make up the chains creating a link between past and present. Only through the 'chains of affinities' can the modifications, resemblances, patterns and descent be seen by the inward eye that Goethe talks about.

This insight enables Darwin to go on to say that he thinks all animals and plants may have descended from one progenitor due to their common structures:

> [...] All living things have much in common, in their chemical composition, their germinal vesicles, their cellular structure, and their laws of growth and reproduction [...]. Therefore I should infer from analogy that probably all the organic beings which

have ever lived on this earth have descended from some one primordial form, into which life was first breathed. (Darwin, 1985, p. 455)

It could be argued that a view holding the position that Darwin was a Romantic biologist does not square with his genealogical 'theory of descent with modifications' in which skulls, vertebrae, jaws and legs have *literally* metamorphosed from a common element, *not metaphorically*. Darwin is keen to emphasize that all forms came from a common element:

> Naturalists, however, use such language only in a metaphorical sense: they are far from meaning that during a long course of descent, primordial organs of any kind – vertebrae in the one case and legs in the other – have actually been modified into skulls or jaws [...]. On my view these terms may be used literally; and the wonderful fact of the jaws, for instance, of a crab retaining numerous characters, which they would probably have retained through inheritance, if they had really been metamorphosed during a long course of descent from true legs, or from some simple appendage, is explained. (Darwin, 1985, p. 419)

However, despite Darwin's insistence that all forms have *literally* metamorphosed from common elements, the archetypes representing the common elements are nevertheless mental constructs enabling Man's understanding of the world, and in particular enabling Darwin to construct his genealogical 'tree of life'. In another sense the archetypes are forward facing and can be seen to be teleological if Darwin's theory of natural selection is seen as a telic enabler. The archetypes are free in that the forms that metamorphose from the common structures over time do so in a non-pre-determined way: they modify themselves according to the need and advantage at the time and at an intermediate step at a time. None of the steps are known in advance. Yet there is a teleological driving force at work within the archetypes ensuring that modifications in form are to the advantage of the organ or species *if survival is to be achieved*. Natural selection is therefore a force enabling archetypes to express themselves through forms that strive towards improvement. This community of form through relatedness can be seen in similar embryonic stages in animals despite their coming from different groups. This demonstrates community of descent from a limited number of progenitors:

> In two groups of animal, however much they may at present differ from each other in structure and habits, if they pass through the same or similar embryonic stages,

we may feel assured that they have both descended from the same or nearly similar parents, and are therefore in that degree closely related. Thus, community in embryonic structure reveals community of descent. (Darwin, 1985, p. 427)

If Darwin's method is seen as 'genetic', this eliminates the contradiction of holding both a genealogical and teleological position at the same time as both can sit on 'the chains of affinities', each being no more than opposite ends of the same continuum, a sliding scale between past and present.

Darwin's imagination can be seen to have been sparked during his voyage on the *Beagle* by the influence of Humboldt's aesthetic method in which he takes his own impressions of Nature into account when relating scientific data and constructing hypotheses. There is undoubtedly an element of Romantic processes running through Darwin's method of viewing Nature that includes an historic perspective that incorporates his own presence as a natural philosopher in the present moment (from a standpoint in which *he* looks back in time and forward into the future in order to comprehend mutability or evolution). Darwin, through his aesthetic imagination, is telling a narrative that includes himself as observer, 'scientist',[23] philosopher and naturalist.

Darwin's concept of natural selection can be seen as Romantic as it uncovers Nature's hidden secret of a self-contained force and set of laws that is forever recreating itself, existing as both creator and product, and existing whether or not there is a God as it exists independently and therefore cannot be subject to God as a first cause. The idea of natural selection is also Romantic as it includes Mankind,[24] and this includes Darwin. It would

---

23   It must also be remembered that Darwin was not only a theorist but a practical scientist, not only collecting and analysing data but also conducting his own experiments whilst on the *Beagle* and in his own house and garden with his children at Down House.

24   As Darwin's *Origin* has shown, Man has evolved from humble physical beginnings, which means that Mind has also evolved from *physical* beginnings in which there was no Mind, no self-conscious reflection. This is Romantic in the sense that when Man reflects on the beauty of Nature, he sees a reflection of himself as he, through the long process of natural selection, has been created by Nature. On the surface, natural selection may appear to be no more than a Malthusian set of self-destructive

be difficult to imagine that Darwin was not aware of these connections when he was obviously struggling to come to terms with the sickness and deaths in his own family and that of his friends, as well as the struggle he was facing with his own decision of whether or not to publish his *Origin*.

Darwin's ability to reflect (in the imaginative sense) on his personal experiences of Nature helped him to learn to 'see' the archetypes in the same way that Goethe did, enabling him, like a bird, to develop the freedom to fly with the flow of the experience of Nature without being constrained by the received Victorian science of the day. Nevertheless, Darwin still lived like a Victorian gentleman, and no doubt this double life contributed to his mental anguish and physical illness.

---

struggles, and therefore not Romantic. But when seen as the law responsible for the creation of Mind, this is truly a mystical form of Romantic alchemy in which opposites can be seen to unite.

CHAPTER 5

Darwin's Concepts of Morality and Romantic Materialism

The mental concept of 'reflection' has already been referred to when considering the importance of relating data and the experiences of Nature in order to grasp its organic whole, and this is particularly important for Darwin when moving between past and present, or when moving between the extremities of the genealogical – teleological continuum showing the descent and development of Nature (its history). The previous chapters have also shown the importance of reflection and imagination through Goethe's 'Genetic Method' when conceptualising the 'leaf' and 'vertebrae' archetypes, enabling the mind to 'see' the commonality running through Nature. This chapter will now continue to examine these concepts of 'reflection' and 'imagination' in relation to Darwin's concept of moral reflection.

The concepts of 'reflection' and 'imagination' gain gradual development from the early Darwin and can be seen as a stronger, consolidated form in the *Descent*. The whole point of the *Origin* and the *Descent* is to show that there is development from animals to humans, and this includes mental as well as physical development. Man has a more advanced form of reflection, but this does not mean that animals have no form of reflection, as Darwin's observation of Jenny the Orang-utan at Regent's Park Zoo showed (referred to earlier). Darwin the natural historian, Darwin the writer and Darwin the moralist when expressing his views on the slave trade, is, through his reflections, demonstrating his advanced mental state as a human being. But all these forms of reflection are building up to the most important one of all for Darwin, that of Man's moral reflection that enables him to tell right from wrong, and enables him to build communities and societies for the benefit of their members. This moral reflection can be seen in Man's feeling of sympathy towards others in pain, or, of remorse when realising that the wrong course of action has been taken. Although it is this moral reflection that sets Man apart from animals, it is at the same time a reflection that over eons of time has developed from animals learning to benefit from mutual cooperation and who have gradually passed this on as an instinct, although Man's instincts are not perfect.

As already discussed, Man processes sense data to understand the objects in Nature but at the same time by reflecting on these objects creates the 'I' or the ego that is doing the reflecting in order to be aware of the self and the world around him. These reflective processes can be seen in Darwin the man and Darwin the natural historian. As a man, like all other human beings, he makes sense of the world through his senses and understands the relationship between himself and the objects in the world by reflecting on his sensations and his thoughts of the world. On a moral level he reflects on his own actions in relation to other beings as well as the actions of others and in so doing forms a moral view, for example against slavery. On a meta level, akin to Humboldt's (aesthetic) Method, Darwin is reflecting on the reflecting process itself and how it evolved from community cooperation in animals and gradually over time became an instinct in humans. This form of 'scientific' reflection, as with Humboldt's Method, can be seen as aesthetic alongside science, poetry, literature, music and art.[1] Each reflection, at whatever level, is an example of Nature creating itself (as already discussed). Just like Descartes' *cogito ergo sum* in which the certainty of the world is founded on the self-consciousness of the 'I':

> That consciousness can be explained in the same terms as the rest of nature is therefore also regarded as *itself* the product of the subject's growing control of nature [...]. German Idealism tries to prove that subject and object are identical, so that the way we think about the world and the world itself are inseparable, because *the world is in fact a subject thinking itself* (emphasis mine). (Bowie, 2003, p. 9)

The chapter will now develop this line of argument with reference to Darwin's *Descent* and his notebooks.

---

[1] This could also include religion, history, philosophy or any other forms of meta reflection that reflect upon our reflections of the world or Nature.

Central to Darwin's concept of Romantic materialism is the notion that Man has developed over eons from simpler animals and those in turn from simpler[2] organisms. The beauty of this form of materialism was that all beings, including Man, came from the same humble beginnings, and that therefore all races and all species ultimately came from the same progenitor, and were and are subject to the same laws of natural and sexual selection. But what sets Man apart from other beings is his heightened sense of awareness of self and others in relation to his environment, his intellect and his moral being which have developed alongside his physical development and have developed from those humble beginnings: 'the mental powers of the higher animals, which are the same in kind with those of man, though so different in degree, are capable of advancement' (Darwin, 2004, p. 679). His intellect enabled him to develop tools to provide shelter and language to interact with other humans as well as the development of social instincts to defend the immediate community. These instincts, Darwin believed, 'have in all probability been acquired through natural selection' (Darwin, 2004, p. 680). This is where Man is different from other beings. Over time he starts to take account of the feelings of others, reflecting on whether his actions receive approval or not, developing his ability to move between reflections on past actions and reflections on present feelings about those actions (the same skill Darwin develops in being able to move between reflections on fossils and their past existences, and reflections on present-day descendants related to their extinct ancestors): 'Man cannot avoid looking both backwards and forwards, and comparing past impressions' (Darwin, 2004, pp. 680–1) and 'Past impressions were compared during their incessant passage through the mind' (Darwin, 2004, p. 123). As Man develops, his actions are governed more by the praise and blame of others rather than 'blind instinctive impulse'. For Darwin, the key to moral development is through sympathy:

---

2   But not necessarily simple in terms of structure as Rebecca Stott highlights in her work on Darwin's study of barnacles: 'This tiny creature had both a life cycle and an adaptation since prehistory [...]; it also had a life history that bizarrely shadowed patterns of human life, shaped as it was by the same natural laws of survival, development and reproduction' (Stott, 2003, p. 246).

> Sympathy, though gained as an instinct, is also much strengthened by exercise or habit. As all men desire their own happiness, praise or blame is bestowed on actions and motives, according as they lead to this end; and as happiness is an essential part of the general good, the greatest-happiness principle[3] indirectly serves as a nearly safe standard of right and wrong. (Darwin, 2004, p. 681)

When looking back at past actions and realising that they were the wrong actions, this 'sense of dissatisfaction' creates a feeling of 'conscience'. This is the cornerstone for determining how future actions ought to be done differently (Darwin, 2004, p. 681).

In advocating the 'greatest-happiness' principle, Darwin was supporting the utilitarian principle of creating the greatest 'general good' for the community. Whilst on the *Beagle* he was horrified[4] at the treatment of slaves, and equally horrified at the notion that 'self-interest will prevent excessive cruelty':[5]

> On the 19th August we finally left the shores of Brazil. I thank God, I shall never again visit a slave-country. To this day, if I hear a distant scream, it recalls with painful vividness my feelings, when passing a house near Pernambuco, I heard the most

---

3 Jeremy Bentham first used this term in his anonymous publication in 1776 *Fragment on Government*, having probably come across the term in Beccaria's *Dei delitti e delle pene*, published in 1768 (Burns, 2005, p. 46). In a passage in his *Constitutional Code*, written in 1822, he writes: 'The right and proper end of government in every political community is the greatest happiness of all the individuals of which it is composed. Say in other words, the greatest happiness of the greatest number. In speaking of the correspondent first principle, call it the greatest-happiness principle' (Burns, 2005, p. 56).

4 As this quote also shows, Darwin's reflective consciousness expressed as both individual and community conscience, can be seen in the choice of words reflecting mental feelings of sympathy: 'painful vividness', 'pitiable moans', 'poor slave', 'tortured', 'powerless as a child to remonstrate', 'cruelty', 'degraded slaves', 'rage ... of masters' and 'protested against with noble feeling'.

5 That is, the slave owners are only refraining from excessive cruelty because sick or injured slaves would reduce their capacity to work and therefore impact on the owners' economic interests. Darwin supports the utilitarians because he wants to see the *whole* community benefit (although, as argued elsewhere in this book, this can be seen to be at odds with his and the Victorians' view that certain members of the community, for example the Irish, should practise 'self-help').

pitiable moans and could not but suspect that some poor slave was being tortured, yet knew that I was as powerless as a child to remonstrate [...]. It is argued that self-interest will prevent excessive cruelty; as if self-interest protected our domestic animals, which are far less likely than degraded slaves, to stir up the rage of their savage masters. It is an argument long since protested against with noble feeling, and strikingly exemplified, by the ever illustrious Humboldt. (Darwin, 1962, pp. 496–7; also cited by Richards, 2002, p. 541)

Darwin supported Mackintosh's[6] objection to Paley's utilitarianism (Richards, 2002, p. 542) in which a human being's moral action was regarded as a response to pleasure and pain in the individual. Against this view, Mackintosh believed that humans acted altruistically for the benefit of others and that this moral sense of right and wrong was part of human nature. Unlike the utilitarian who defined the right action according to how much pleasure or pain it would produce,[7] Mackintosh believed that humans just knew innately what the right action was. Although Darwin was hugely influenced by Mackintosh, he was critical of him providing no explanation as to the origin of the 'moral sense';[8] that is, what enabled a person to know instinctively that one action was right and another wrong

6   Darwin read his *Dissertation on the Progress of Ethical Philosophy* (1837, second edition).
7   According to Bentham's *Introduction to the Principles of Morals and Legislation* (1789), the amount of pleasure or pain produced by an action could be calculated using his 'felicific calculus' which took into account such factors as the pleasure or pain's 'intensity', and 'duration'. Bentham's philosophy was based on the premise that mankind's behaviour was governed by the desire for pleasure and the desire for the absence of pain. A right action was defined as that which increases pleasure, and a wrong action as one that decreases pleasure. But the benefit created is the one that benefits all members of the community even though not all members are pleased equally. Bentham does not elevate the community above the individuals (Zunjic, 2014, unnumbered page).
8   The 'moral sense' theory came from Frances Hutcheson who believed that the 'moral sense' was implanted by God. His theory is expounded in *An Enquiry Concerning the Original of our Ideas if Virtue or Moral Good* (1725) and *An Essay on the Nature and Conduct of the Passions and Affections, with Illustrations of the Moral Sense* (1728) (Vandenberg and DeHart, 2014, unnumbered page). James Mill attacked Mackintosh's version of the 'moral sense' because it could be seen to override the

(Richards, 2002, p. 543). For Darwin habits became instinctive after being practised over a number of generations. Those instincts passed on to the next generation that helped preserve family and community groups could be regarded as moral, whereas those desires that only concerned individuals could not.[9]

The feeling of sympathy, derived from habit and then instinct, can be seen to be the central plank of both Man's and Darwin's sense of morality; in terms of how one relates to others within the immediate community, society, as well as the treatment of one's environment including other species, and one's self in terms of how one views one's moral conduct in relation to the world and its inhabitants. This moral view of oneself and others in society, in terms of sympathy and conscience, would include attitudes towards slavery, treatment of workers in the towns and on the land, both at home and in Ireland and the colonies. In another sense the morality reflecting an era, its sympathies and conscience, can be said to evolve. Darwin's time, however, did not consist of one set of beliefs as the era was made up of different classes with different views of what was right and wrong (for example, voting rights and fair pay) and therefore their sympathies and consciences were often in conflict with each other. Nevertheless Darwin believed morality was progressive and that Man in civilized nations could achieve a high moral state 'through the advancement of his reasoning powers and consequently of a just public opinion [...] through the effects of habit, example, instruction, and reflection', conscience then becoming 'the supreme judge and monitor' (Darwin, 2004, p. 682). Darwin believed these instincts were 'primarily gained [...] through natural selection' (Darwin, 2004, p. 682).

The notions of sympathy and conscience as outlined above could be interpreted as Romantic in derivation because Darwin sees these reflective

---

judgement of utility: if utility was the moral judgement then the 'moral sense' would have to be seen as immoral rather than moral (Coplestone, 1967, p. 36).

9   Darwin's explanation of how instinct develops into moral behaviour is not fully worked out despite Richards' belief that Darwin had 'found a biological explanation' (Richards, 2002, pp. 545–6). But this is perhaps understandable considering the limited knowledge of inheritance that was available to Darwin at the time.

mental constructs as having developed over time through natural selection and, most importantly, having emanated initially through habit before becoming an ingrained instinct. These habits developed originally from habits that were non-reflective 'simpler instinctive actions' that existed before the development of a human brain that could think, before it had an intellect, and therefore material:

> Some intelligent actions, after being performed during several generations, become converted into instincts and are inherited, as when birds on oceanic islands learn to avoid man [...]. But the greater number of the more complex instincts appear to have been gained in a wholly different manner, through the natural selection of variations of *simpler instinctive actions*. Such variations appear to arise from the same unknown causes acting on the cerebral organisation, which induce slight variations or individual differences in other parts of the body; and these variations [...] are often said to arise spontaneously (emphasis mine). (Darwin, 2004, p. 88)

Darwin admits that these changes arise from 'unknown causes' but feels confident that there is a link between repetition (habit), which through continued use, somehow makes changes in the brain to bring about instinct. It is this mysterious notion of the Mind having been generated by natural laws from Nature that enables it to be regarded as Romantic. Darwin recognizes that as primitive Man develops over time in using tools and in forming the first words of language for communication, these 'variations of simpler instinctive actions' act 'on the cerebral organisation', that is, they effect changes in the brain. This means therefore that as all creatures, including Man, developed originally from simple unthinking organisms, the mind developed from the physical being. In a nutshell, Man came from Nature, was created by Nature, and is therefore very much a part of Nature:

> When primeval man first used flint-stones for any purpose, he would have accidentally splintered them, and would then have used the sharp fragments. From this step it would be a small one to break the flints on purpose, and not a very wide step to fashion them rudely [...]. The orang is known to cover itself at night with the leaves of the Pandanus; and Brehm states that one of his baboons used to protect itself from the heat of the sun by throwing a straw-mat over its head. In these several habits, we probably see the first steps towards some of the simpler arts, such as rude architecture and dress, as they arose amongst the early progenitors of man. (Darwin, 2004, p. 104)

The mental development of the brain can be seen in the behaviour of primitive Man in the concept of 'purpose' and 'fashioning', and by analogy in the orang in 'protecting' itself from the heat of the sun by using the mat as a tool. By using the orang as an example, it gives the reader an idea of how Man might have developed from an earlier branch of his own tree of descent.

The *Descent* was published in 1871, but Darwin had been thinking about these metaphysical mind-body issues for many years before and these thoughts can be clearly seen in his *Notebook M*,[10] *Notebook N*,[11] and his *Old & Useless Notes*[12] written between 1838 and 1840 (transcribed and edited in Barrett, 2008). In *Notebook N* these early references to conscience and instinct can be seen in '[...] any animal with social & sexual instinct <<& yet with passion>>[13] he *must* have conscience [...]. Dogs [sic] conscience would not have been same with mans [sic] because original instincts different' (Barrett, 2008, p. 564). Here Darwin identifies the level of conscience with type of instinct. The type or level of conscience can be seen to be reflected in his developed view of morality in which individuals and society can be graded according to their developed moral view with the most 'civilized' being regarded as those that have progressed the most. This embryonic view in its simplest form can be seen in his 'different nations having different moral sense [...]' and '[...] man moreover who *reasons* much on his actions, makes his conscience far more sensitive', the 'more sensitive' being a moral sense that could be interpreted as being more superior (Barrett, 2008, p. 564).

Also in *Notebook N*, the influence of habit is identified as having an effect on the brain: 'an habitual action must some way affect the brain in

---

10 This 'is the first of several sets of notes [...] on the general subject of the biological origin of behaviour' written between July 15–12 October 1838 (Barrett, 2008, p. 517).
11 This 'is the second notebook on "metaphysical enquiries" and expression', the first page dated 'October 2d. 1838' and the last dated entry '28 April 1840' (Barrett, 2008, p. 561).
12 This notebook is not actually a notebook 'but a collection of notes on miscellaneous sheets of paper [...]. The title is misleading in that it reflects the dismissive attitude of an author filing away notes no longer useful [...]' (Barrett, 2008, p. 597).
13 The symbol <<>> means Darwin's insertion. See 'Symbols used in the transcriptions of Darwin's notebooks', Barrett, 2008, p. vi.

a manner which can be transmitted' (Barrett, 2008, p. 574). But Darwin also wants to emphasize his theory that habitual actions lead to instincts through natural selection; that is, those instincts which improve a species' prospects of survival will give the species an advantage over other species, or different populations of the same species in the form of tribes or races.[14] Darwin argues by analogy, using the strong arms of a blacksmith being passed down to his children as an example:

> An habitual action must some way affect the brain in a manner which can be transmitted. – This is analogous to a blacksmith having children with strong arms. – The other principle of those children. Which *chance*? Produced with strong arms, outliving the weaker ones, may be applicable to the formation of instincts, independently of habits. – the limits of these two actions either on form or brain very hard to define. (Barrett, 2008, p. 574)[15]

Although Darwin is here using his own theory of natural selection, his example of the blacksmith can be seen to be Lamarckian[16] in terms of

14   'Extinction follows chiefly from the competition of tribe with tribe, and race with race. Various checks are always in action, serving to keep down the numbers of each savage tribe – such as periodic famines, nomadic habits and the consequent deaths of infants, prolonged suckling, wars, accidents, sickness, licentiousness, the stealing of women, infanticide, and especially lessened fertility' (Darwin, 2004, p. 212). This fits in with his idea of the more 'civilized' races 'exterminating' the more 'savage' races. Those that have the more 'civilized' qualities have a more selective advantage over those that do not.
15   It is important to note that this entry in Darwin's *Notebook N* was written sometime between October 1838 and April 1840 (Barrett, 2008, p. 561). It is understandable, therefore, that his views at this time should lean heavily towards Lamarck's as his own theory of natural selection was still very embryonic in form. Darwin referred to this notebook in December 1856 for work on the *Origin* and, in May 1873, for the *Descent* and *The Expression of the Emotions in Man and Animals* (Barratt, 2008, pp. 562–3).
16   Darwin was not consistently Darwinian in that in the *Origin* he allowed Lamarckian inheritance of acquired characteristics to run alongside his theory of natural selection offering an additional mechanism of biological evolution. Darwin called this view the 'Pangenesis hypothesis' (Darwin, 2004, p. 264). This theory proposed that each organ passed information on how they were used or not used to sperm and eggs via 'gemmules'. This was done throughout an individual's life, passing information on

development being brought about by constant use but the difference here is that change, or in this case instinct, is passed on through inheritance[17] rather than the constant use of a particular part of the body. The main issue here is that Darwin wants to show that an 'habitual action [...] affects the brain' thus underpinning his view that habits lead to moral instincts or moral sense, but not input by God but created by Nature. Darwin is also making the point that these instincts have not been brought about by *chance* but have been brought about by the law of natural selection created by Nature. His anthropomorphic use of 'strong [human] arms' to argue the case by analogy also helps draw in the human form of mind and body into the fold of Nature, and by association makes morality part of Nature.

The mind-brain issue is again considered by Darwin when he examines the cross of instincts produced when crossing jackals, foxes, wolves and dogs. Darwin asks 'can we deny that brain would be intermediate like rest of body?' (Barrett, 2008, p. 575). In other words, can the brain adapt to its environment and change in the same way as species adapt and change to become incipient species or varieties (that is, 'intermediate' between two forms)? Not only does Darwin concede that there must be some relationship between the mind and the brain but he also believes that there is no difference between the two: 'can we deny relation of mind and brain. <<Do we deny the mind of a greyhound & spaniel. differs from their brains>>[18] (Barrett, 2008, p. 575). Darwin also feels that the instinct of digging for mice passed down from the dog to its pup offspring could

---

to the next generation. So for example, men with big muscles used often would pass this on to their children (Hurley, 2009, unnumbered web page).

17   It was not until after Darwin's death in 1883 that August Weismann introduced his 'Germ-Plasm' Theory, in his lecture *Über die Verebung* (On Inheritance), which finally laid to rest Darwin's Pangenesis theory. Weismann's theory stated that inheritance only took place by means of the gametes or germ cells (the egg and sperm cells). The other cells of the body (somatic cells) did not function as agents of heredity as inheritance only worked by going in one direction from the gametes to the somatic cells and not back from the somatic cells to the gametes. Acquired characteristics could not therefore be inherited (Hurley, 2009, unnumbered page).

18   The symbol <<>> means the text has been inserted at a later date by Darwin (see 'symbols used' in Barrett, 2008, p. vi).

come 'from some peculiarity of structure of brain', and asks 'can we suppose some *essence*' (emphasis mine) (Barrett, 2008, p. 575). This suggests that Darwin is seriously considering an archetype for the mind-brain analogous to the vertebrae and leaf archetypes discussed earlier. Although the mental concept of 'archetype' or 'essence' is at a higher level of mental reflection than the mental processing of physical and mental sensations, Darwin sees mental pain (grief) as the same as physical pain since 'tears flow from both' and that the same applies to the 'pleasure of senses' (Barrett, 2008, p. 575).

As mentioned earlier, Mackintosh influenced Darwin in the development of the theory that morality came from instincts (although Darwin was critical of Mackintosh in not explaining how these instincts were developed and where they resided). In *Notebook N*, Darwin refers to 'Mackintosh Ethics' and his 'Theory of Association [...] when [ideas] entered brain' (Barrett, 2008, p. 575). Mackintosh was here referring to Berkeley and Hume's agreement that the intellectual operations of the mind were due to the *association of ideas*.[19] But for Darwin these 'ideas' are part of the physical make-up of the brain, and are not independent of the brain, which is why he says 'try contiguity of parts of Brain' (Barrett, 2008, p. 575). This physical location of ideas (or for Darwin, instincts) in the brain is emphasized in Darwin's own marginalia in Mackintosh's book referring to the *association of ideas*: 'try theory of place in brain' (Barrett, 2008, footnote 4, p. 587). Yet although the mind, the brain and the self can be seen as one, Darwin believes he has identified different layers of consciousness, or 'double consciousness', which separates the self or will from imagination. When referring to the 'peculiarity of structure of brain' he asks if this can be 'double consciousness' (Barrett, 2008, p. 575). Beer traces this influence to his reading of the Romantic poets when she states 'he was intensely aware, perhaps in part through Coleridge, of the multiple voices of consciousness'

---

19   'Both agree in referring all the intellectual operations to the *association of ideas*, and in representing that association as reducible to the single law, that ideas which enter the mind at the same time, acquire a tendency to call up each other, which is in direct proportion to the frequency of their having entered together' (Mackintosh, 1837, p. 248). Also cited by Barrett, 2008, footnote 4, p. 587).

(Beer, 2010, p. 5). In this context Beer also refers to Darwin's use of the term 'double consciousness' in his *Notebook N*:

> Double consciousness only extreme step of an ideal argument held in one's own mind, & Dr Hollands story of man in Delirium tremens hearing other man speaks. [sic] shows, that consciousness of personnal [sic] identity is by no means a necessary part of man's mind. – (Barrett, 2008, p. 593)

Beer sees this 'double consciousness' within Darwin's own thought processes. When reading, he engages in a 'conversation' (even regarding it as a kind of music) or an 'active silent dialogue in which the reader slides into the place of the writer and yet presses back into his or her own person' (Beer, 2010, pp. 5–10). This dialogue of a 'double consciousness' can be seen to reflect both Goethe's 'Genetic Method' and Humboldt's aesthetic method in which the imagination is given free rein to move between the particular and the universal, between the subjective observer and the observed, between empirically based fact and the imaginative possible. This 'double consciousness' conversation is also very Wordsworthian and can be seen in Wordsworth's poem *The Excursion*, which Darwin read several times:[20]

> His reading of *The Excursion* prompted a parallel vivid 'double consciousness'. The *Excursion* is itself a poem of conversation,[21] ethical and metaphysical debate, and the telling of tales. It frames the lives of others through the recollections of those

---

20 Reference to Darwin's reading of Wordsworth's poetical works can be found *at Darwin Correspondence Project* (2014a, p. 164). Also, in his *Autobiography*, he states that he 'can boast that I read the *Excursion* twice through' (Darwin, F., 1995, p. 31).

21 The conversations in the poem are 'concerned with Wordsworth's personal conflicts, sorrows, and consolations, thinly disguised in the dialogue between his Wanderer (the "Pedlar" of *The Ruined Cottage*), the despondent Solitary, and the pious old village priest'. In one part of the poem where the Solitary sees a reflection of himself in a clear mountain stream, 'we see the very image of Wordsworth's soul, successively turbulent and peaceful; bearing even in repose the signs of former conflicts; a soul driven indeed by passion yet not enslaved by it; and, when stillness comes, reflecting in its depths all things in heaven and earth'. (Moorman, 1965, pp. 77–80). This inner turbulence or mental struggle, and 'a soul driven indeed by passion yet not enslaved by it', can be seen to be similar to Darwin's short-lived passions and more permanent moral instincts (referred to earlier in his *OUN*).

who have observed them. The philosophical musings arise out of homely instances. (Beer, 2010, p. 8)

The 'homely instances' is key here, as, for Darwin, slight changes over time can 'produce great transformation'. Nothing was too trivial for Darwin – the life of a beetle, a worm or a barnacle could unlock the secrets to the universe. The 'mind's *excursive* power' (Wordsworth, 1949, p. 119) enables the mind to move between science and empiricism on the one hand, and to imaginative 'castles in the air' (Barrett, 2008, p. 527) to form theories on the other. But perhaps most importantly, this ability of the mind to move from the ordinary to an enlarged insight of Nature, enabled Darwin to trace the morality of Man from humble physical beginnings (Beer, 2010, p. 8).[22] This 'double consciousness' conversation of the voices in the poem, or, in Darwin's case, the prose of his text, can also be seen as a textual effect of the form of writing which is 'self-conscious'.[23] Although the poet or writer creates the meaning, the poem or piece of prose also 'has a mind (and an unconscious) of its own' (O'Neill, 1997, xv). Such self-consciousness is common among the Romantic poets as they 'put in question the idea of a knowable or discoverable self', the imaginative self resting somewhere between the self made up of 'a bundle or collection of different perceptions' and an absolute self experiencing 'sublime moments' (O'Neill, 1997, p. xvii). There is frequently such self-conscious searching for knowledge

---

[22] Beer cites Robert Richards (1989) as placing emphasis on 'the degree to which Darwin sought to find a particular place for human morality in his argument' (Beer, 2010, p. 8).
[23] O'Neill defines 'self-consciousness' (of a poem) 'as textual effects, products of a text's procedures that may persuade the reader that they are, at least partly, agents involved in the process of the text's production'. The self-conscious form of the poem tells the reader 'what the poem is saying or finding out about itself – or not saying and not finding out about itself' (O'Neill, 1997, p. xiv). It is also about 'becoming' as perfection in the quest for knowledge is never totally achieved (O'Neill, 1997, p. xx).

in the textual forms of the *Origin*[24] and the *Descent*[25] through the use of conversational questions. This quest may be seen as Romantic because the texts of such writing reveal a mixture 'of self-assertion and self-doubt; they know of no mode of knowledge superior to their own, yet their sense of a contract with an audience is fragile' (O'Neill, 1997, p. 36). In parallel to the development of Nature that Darwin is interrogating, Darwin's own quest for the knowledge of Nature can also be seen as a form of 'becoming'.[26] In Wordsworth's *The Excursion*, which deals with the concepts of time and transience, such self-consciousness enables the poet to get nearer to experiencing that which cannot be experienced, namely death. As pointed out by O'Neill,[27] '*The Excursion*'s awareness of itself as a poem is a means of bringing death and temporality within "the reach of reflection"' (O'Neill, 1997, p. xxxviii). With Darwin, this same type of self-conscious awareness enables him to explore the concepts of extinction and death as well as the existence of Man's progenitors before the life of Man. Another aspect of self-consciousness that flows through the writings of both Wordsworth and Darwin is that 'imagination' is linked with 'morals'. For Darwin his imaginative quest for knowledge seeks to discover the physical causes of morality in Nature, but at the same time believing in a social order that supports the equality of races in terms of their common progenitor (although the

24 For example, in the chapter on natural selection in the *Origin*, the use of questions such as 'Can the principle of selection, which we have seen is so potent in the hands of man, apply in nature?' and 'Can it, then, be thought improbable, seeing that variations useful to man have undoubtedly occurred, that other variations useful in some way to each being in the great and complex battle of life, should sometimes occur in the course of thousands of generations?' (Darwin, 1985, p. 130).

25 For example, in the chapter on the moral sense in the *Descent*, the use of such questions as 'Why should a man feel that he ought to obey one instinctive desire rather than another? Why is he bitterly regretful, if he has yielded to a strong sense of self-preservation, and has not risked his life to save that of a fellow-creature? Or why does he regret having stolen food from hunger?' (Darwin, 2004, p. 134).

26 O'Neill, referring to Friedrich Schlegel's account of Romantic poetry in the *Atheneäum Fragments (1798)*, sees the self-conscious poem as being associated with 'becoming' as it is forever seeking perfection without attaining it (O'Neill, 1997, p. xx).

27 O'Neill (1997, footnote 51, p. xxxviii) cites Paul de Man's discussion of this point (Paul de Man, 1993, p. 63).

races may be at different stages of development). For Wordsworth, he 'is alive to the way language involves speakers and listeners in struggle about value', although he 'is haunted by the fear that some modes of visionary imagination are resistant to moral approbation' (O'Neill, 1997, p. 33).

For the young Darwin poetry and music played an important part in sparking his imagination (or 'castles in the air') and the development of his concept of moral reflection. They cause 'the mind to create short vivid flashes of images & thoughts. – Poetry. [sic] the latter thoughts are in same manner vivid & grand' (Barrett, 2008, p. 527). Perhaps Darwin's dislike for poetry[28] in later years was a reflection of his 'fossilized' scientific mind making it difficult for him to create 'castles in the air' any more. Perhaps his scientific ideas were so well worked out that he did not have the need to stretch his imagination to the same extent to think of impossible dreams not grounded in empirical science. But the younger Darwin, freshly back from his *Voyage* is very much in need of them, and it is these 'castles' of stretched imagination that lay the groundwork for both the *Origin and* the *Descent*:

> Granny[29] says she never builds castles in the air. Catherine[30] often, but not of an inventive class. – Now that I have a test of hardness of thought, from weakness of my stomach I observe a long castle in the air, is as hard work (abstracting it being done in open air, with exercise &c no organs of sense being required) as the closest train of geological thought. – the capability of such trains of thought makes a discoverer, & therefore (independent of improving powers of invention) such castles in the air are highly advantageous, before real train of inventive thoughts are brought into play & then perhaps the sooner castles in the air are banished the better. – (Barrett, 2008, p. 527)

---

28 In Darwin's *Autobiography*, published posthumously by his son Francis in 1892, he states 'But now for many years I cannot endure to read a line of poetry' (Darwin, F., 1995, p. 50). The unpublished work had the heading 'Recollections of the Development of my Mind and Character' and was dated 'Aug. 3, 1876' (Darwin, F., p. 5). Darwin wrote this when he was sixty-seven years old, six years before he died in 1882.
29 'Granny' refers to Darwin's sister Susan, one of his older sisters, who acted as his mother after his mother's death when a child.
30 His sister Emily Catherine (known as Catherine, Catty and Kitty).

For Darwin, 'castles' were a form of imagination that 'invented' ideas that helped create discoveries. The 'sublime' could also be seen as an aid to this. There are many theories surrounding Darwin's chronic illness (stomach upsets, fevers, blisters, etc.) which he suffered on and off for most of his adult life, but it is likely that the stress of living with his controversial 'dangerous idea' of evolution contributed to this. Discovering the 'tree of life' and peering back at it along its apparent infinite chain must have seemed like standing on the edge of an abyss, and this could have been a sublime experience creating mixed feelings of awe, fear and terror. Yet in order not to be swallowed up by it, he had to create a 'hardness of thought' to overcome his 'weakness of stomach'. His notes here read almost like a poem with a bundle of contradictory statements. He builds his 'castle' from geological data he has gained from experience in the open air, but the relating of the data to form the abstraction is done without the help of his sense organs. This freedom to construct such 'castles' gives him independence from the received view of science, but, at the same time he has to use scientific hypotheses to test out his ideas. The 'castles' are then tempered by scientific reason to create the 'real train of inventive thoughts'. For Darwin these 'castles' need to be banished by 'inventive thoughts', the science with a breathed-in type of Wordsworthian poetry. This banishment is

> The facility with which a castle in the air is interrupted & utterly forgotten -, so as to feel a severe disappointment <<in real train of thought this does not happen. because papers, &c &c round one. one recalls the castle by going to beginning of castle>> because train cannot be discovered – is closely analogous to my Fathers [sic] positive statement that insanity is only cured by forgetfulness. – & the approach to believing a vivid castle in the air, or dreams real again explains insanity. – (Barrett, 2008, pp. 527–8)

The 'castles' of imagination are so far removed from reality as to be akin to 'dreams' or 'insanity'. 'Insanity' could be seen as an extreme form of the sublime in which one is lost in imaginative dreams, or 'castles', and is unable to come back to reality. The 'train of inventive thoughts' already have a literal paper trail of notes to work out the beginning and the end, to work out the order of reasoning, enabling interruptions to occur without destroying the structure. 'Castles' on the other hand are not grounded in reality, are more ephemeral and more susceptible to interruptions and therefore are more easily lost or forgotten. This movement or 'conversation' between 'castles in the air' and 'train of inventive thoughts' is akin to the movement between bodily sensations and thoughts in the brain or mind, as well as Humboldt's aesthetic Method.

In his *Notebook M*, Darwin discusses the way the imagination can move between poetry, art, music and science, referring to Wordsworth's (1801/1802) Preface to *Lyrical Ballads* in which the poet hopes that when science has been sufficiently absorbed into the community it will become truly 'poetic' (Beer, 2010, p. 5):

> Pleasure of imagination [...] connection with poetry. I a geologist have illdefined [sic] notion of land covered with ocean, former animals, slow force cracking surface &c truly poetical. (V. Wordsworth about science being sufficiently *habitual* to become poetical) (emphasis mine). (Barrett, 2008, pp. 528–9)

Wordsworth's 'science being sufficiently habitual to become poetical' referred to by Darwin[31] can be found in the following passage from the Preface to *Lyrical Ballads*:

> Poetry is the first and last of all knowledge – it is as immortal as the heart of man. If the labours of Men of Science should ever create any *material revolution*, direct or indirect, in our condition, and in the impressions which we *habitually* receive, the Poet will sleep then no more than at present, but he will be ready to follow the steps

---

31   It is important to note that Darwin's geologist mentor, Sedgwick, was a friend of Wordsworth's and they went on geological walks in the Lake District together. This friendship is demonstrated in Sedgwick's letters to Wordsworth: 'Some of the happiest summers of my life were passed among the Cumbrian mountains and some of the brightest days of those summers were spent in your society and guidance' (Clark and Hughes, 1890, I, 248–9, cited by Gaul, 1979, pp. 34–5). In addition to being a Geology Professor at Cambridge, he was fond of literature and encouraged the young Darwin 'to read Shakespeare, Milton and Wordsworth, in the years from 1837 to 1839, the years Darwin was working on the Metaphysical Notebooks' (Gaull, 1979, pp. 34–6). Sedgwick was therefore an important link between Darwin and Wordsworth and between science and poetry. Geology, both scientifically and poetically, was the lynchpin to Darwin's theory of natural selection.

of the Man of Science, not only in those general indirect effects, but he will be at his side, *carrying sensation into the midst of the objects of the Science itself* (emphasis mine). (Wordsworth, 2003, pp. 16–17)

Darwin's reference to the word 'habitual' is significant here in the context of aesthetics as his argument in the *Descent* wants to demonstrate that habit creates instinct. Although here he is talking about an aesthetic sense being developed through the absorption of science by the community, this higher aesthetic sense can also be seen as a moral sense (that is, science, poetry, art or music being developed for the good of the community). A *'material revolution'* created by science need not be feared if the Poet is there to breathe in the sensations of Nature into its midst.

Stating that biological behavioural changes came about through generations of habit becoming instinctive, enabled Darwin to adopt a 'slightly altered Lamarckianism' (Herbert and Barrett, 2008, p. 518). Whereas Lamarck believed 'that organisms acted consciously in altering their behaviour' (Herbert and Barrett, 2008, p. 518), Darwin wanted to show that although the mind affected the body, this was done *unconsciously from habit or instinct*. Evidence of this can be seen in Darwin's *Notebook C*[32] in which he says 'Lamark's [sic] willing absurd, [therefore] not applicable to plant' (Barrett, p. 259); his reference to structures being instinctive and hereditary: 'heredetary [sic] ambling horses, (if not looked at as instinctive) then must be owing to heredetary [sic] power of Muscles. – then we SEE structure gained by habit' (Barrett, 2008, p. 290). Again, in Man's case, he shows that the active use of memory (or will) is not required when something becomes instinctive:

> My view of instinct explains its loss ſ [sic] if it explains its acquirement. – Analogy. A bird can swim without being web footed yet with much practice & led on by circumstanc [sic] it becomes web footed, now Man by effort of Memory can remember how to swim after having once learnt, & if that was a regular contingency the brain would become webfooted [sic] & there would be no act of memory. – (Barrett, p. 293)

---

32   *Notebook C* was written between February and July 1838.

Darwin emphasizes this view that habits and habitual instincts precede structure (the fixed instinct) and that this is not brought about by a conscious mental state or will: 'according to my view, habits give structure, ... habits precedes [sic] structure, ... habitual instincts precede structure. – duckling runs to water. Before it is conscious of web feet.-' (Barrett, p. 301).

This view that the unconscious structure of instincts resides somewhere in the brain could be regarded as materialist, and this is the view taken by Herbert and Barrett when referring to *Notebook M*: 'The most obvious sign of Darwin's new perspective was the alignment of his transmutation view with materialism and determinism. He embraced materialism enthusiastically [...] and argued [...] that thought originated in sensation' (Herbert and Barrett, 2008, p. 519). For example:

> Fear must be simple instinctive feeling; I have awakened in the night, being slightly unwell & felt so much afraid though my reason was laughing & told me there was nothing [...]. The sensation of fear is accompanied by <<troubled>> beating of heart, sweat, trembling of muscles. (Barrett, 2008, p. 532)

However, as argued in earlier chapters, this form of materialism should be regarded as 'Romantic materialism' (Beer, 1983, p. 152) as the underlying cause is more of a metaphysical concept than a tangible physical entity, despite being part of a material brain (thus the apparent oxymoron 'Romantic materialism'). At the time of writing *Notebook M* Darwin was visiting his father,[33] a doctor, and discussed various medical conditions, including depression, epilepsy, hereditary defects which helped inform his ideas. Darwin refers to states of mind being brought about by sensations or 'bodily causes':

> Ill-humour & depression, which comes on from bodily causes. -. It is an *argument for materialism*. that cold water brings on suddenly in head, a frame of mind, analogous to those feelings, which may be considered as *truly spiritual* (emphasis mine). (Barrett, 2008, p. 524)

---

33   Robert Waring Darwin. There are frequent references to his father such as 'My F. says', 'My father's test of sincerity', 'My father thinks', 'My father says', 'My father does not know whether ...' (Barrett, 2008, pp. 524–50).

The seemingly opposing terms 'materialism' and 'truly spiritual' show the apparently contradictory nature of Mind and Brain, and therefore the term metaphysical[34] is apt. This form of materialism made him doubt the existence of free will (seeing it as the same as chance):

> One doubts existence of free will every action determined by heredetary [sic] constitution [...]. I verily believe free-will & chance are synonymous. – Shake ten thousand grains of sand together & one will be uppermost: – so in thoughts, one will rise according to law. (Barrett, 2008, pp. 526–7)

Chance does not mean that there is no causation only that the cause is not known. Similarly, 'the illusion of free will is likewise only an illusion that there is no causal necessitation of the feeling, belief or decision enacted by the mind' (Hodge, 2003, pp. 56–7). As Darwin subsumed mind under matter causally early on in his notebooks in 1838, he 'never later had to construct new ways to secure the continuity between man and animals or between man and the lawful order of nature' (Hodge, 2003, p. 57).

To an extent, therefore, according to Darwin, thoughts are determined 'according to law', shaped by a brain that has developed and adapted itself to its environment through the law of natural selection. In *Notebook M*'s discussion of the Mind-Brain problem, one can see Darwin wrestling with the issue of morality that he develops in the *Descent*. Here Darwin already sees 'conscience' as 'instinct', 'acquired by senses', and that therefore 'thinking consists of sensation of images'. The 'instinctive knowledge' of those sensations which brings about those feelings of 'conscience' is the 'memory of such sensations, & memory is repetition of whatever takes place in brain. [sic] when sensation is perceived' (Barrett, 2008, p. 534).

To fully understand Darwin's 'reflective' thinking and imagination, and his concept of morality, it is important to examine the relationship between the notions of sympathy and the sublime which Darwin discusses in his *Old*

---

34 Darwin regarded the subject matter in *Notebook M* as 'metaphysical': 'In his 'Journal' begun in August 1838, Darwin referred to [...] *Notebook M* as follows – 'opened note book connected with Metaphysical Enquiries', cited by Herbert and Kohn, 2008, p. 7, footnote 2 (Herbert & Kohn, 2008, p. 7).

and Useless Notebooks (OUN).[35] His notes discuss Dugald Stewart's *Essay on the Sublime* (Stewart, 1829) in which Longinus[36] remarks that the sublime 'fills the reader with a glorying, and sense of inward greatness' (Stewart, 1829, p. 268). Longinus wrote *On the Sublime* which is an aesthetic and ethical dissertation on a style of writing (for example, by Homer or Plato) in which the writer and reader can form an empathy that goes above the ordinary form of writing, taking both to a level of ecstasy in which greatness of soul is achieved. The sublime can be experienced not only through beauty, but also through that which is distressing causing bewilderment. As referred to by Darwin:

> The emotions of terror & wonder so often concomitant with sublime. adds not a little to the effect: as when we look at the vast ocean from any height. – That the superiority & 'inward glorying', which height. [sic] by its accompanying & associated sensations so often gives. (Barrett, 2008, p. 605)

Stewart conveys the meaning of the sublime by showing how the mind experiences the difference between downward motions and upward motions. A bird soaring upwards, for example, 'exhibits *active powers* which are completely denied to ourselves', and this gives us the idea of a supernatural agent' (Stewart, 1829, p. 270) which is what we mean when we talk of 'flights of imagination and of fancy' (akin to Darwin's 'castles in the air'). Such 'powers are commonly supposed to be the immediate gift of heaven; and not like our scientific habits [and education]' (Stewart, 1829, pp. 270–71). Darwin refers to 'great height' and 'eternity' creating this feeling of an 'inward glorying' which is the sublime:

> Hence it appears, that when certain causes, as great height, eternity, &c &c. produces an inward pride & glorying. [sic] (often however accompanied with terror &

---

35 Barrett, 2008.
36 His work *On the Sublime* (First Century AD) was 'the first time greatness in literature is ascribed to qualities innate in the writer rather than in the art' (*Encyclopaedia Britannica*, 2014, unnumbered web page).

wonderment), <which>[37] <<this>> emotion, from the associations before mentioned. [sic] we call sublime. (Barrett, 2008, p. 605)

This captures an image of the imagination moving between the height of the sublime and the humbleness of the ordinary (yet the outward appearance of humbleness can often hide a sublime level of complexity, as with the barnacle). This can be seen not only in Darwin's jumping from the detail of a beetle in one sentence to an examination of the law of natural selection in the next, but also in the switching between the empirical and the scientific (or as already quoted, between 'heaven' and 'our scientific habits'). The 'flights of imagination' that enabled Darwin to develop his insights into Nature come from 'the idea of Creative Power, [which] is owing, in part to the irresistible tendency which that idea has to raise the thoughts toward Heaven' (Stewart, 1829, p. 283). The feeling of sympathy (an aspect of 'moral sense') is created through one's imagination making one's feelings go out towards the object, in which the subject and object become one:

> It appears to me, that we may often trace the source of this 'inward glorying' to the greatness of an object itself or to the ideas excited & associated with it. [sic] as the idea of Deity. with vastness of Eternity. [sic] which superiority we transfer to ourselves in the same manner as we are acted on by sympathy. (Barrett, 2008, p. 605)

This feeling of sympathy plays an important part in Wordsworth's subjective experience of Nature in which that experience is actually communicated to the reader and is therefore objectified, such that it can be said to exist in both the poet's and reader's mind:

> I have said that Poetry is the spontaneous overflow of powerful feelings: it takes its origin from emotion recollected in tranquillity: the emotion is contemplated till by a species of reaction the tranquillity gradually disappears, and an emotion, kindred to that which was before the subject of contemplation, is gradually produced, and *does itself actually exist in the mind* (emphasis mine). (Wordsworth, 2003, p. 21)[38]

---

37   The symbol <> refers to an item deleted by Darwin (see symbols used in the transcriptions of Darwin's notebooks, Barrett, 2008, p. vi).
38   In Preface to *Lyrical Ballads*.

An example of this communication of the subjective experience of Nature and the emotional experience of sympathy that accompanies it can be seen in Wordsworth's *The Prelude* in which he retells the experience of himself as a boy taking a boat out without permission on Ullswater lake. But in this poem he is not just conveying the experience of the boy to the reader but also the poet's feeling of the way in which Nature can penetrate the observer's mind, and, in so doing, become objectified:

> Wordsworth so describes the boy's experiences as to recreate in the reader sensations and feelings, the 'emotion', experienced by the boy. But the action on the reader of this sequence of feelings is to generate a new emotion, akin to but different from that consciously felt by the boy. It is this complex emotion which, experienced by the reader, constitutes the poet's communication of the 'influence of natural objects'. (Winkler, 1975, p. 159)

Both the boy and the reader of the poem are shocked by the sudden change from the movement of the boat 'like a Swan' to the appearance of 'a huge cliff' with its 'uprear'd head' like an unearthly creature instilling sublime fear:

> I dipped my oars into the silent lake,
> And, as I rose upon the stroke, my boat
> Went heaving through the water, like a Swan;
> When from behind that craggy steep, till then
> The bound of the horizon, a huge Cliff,
> As if with voluntary power instinct,
> Uprear'd its head. (Wordsworth, 1970, lines 402–8)

The fear of the sublime is emphasized with such phrases as 'that spectacle', 'undetermined sense /Of unknown modes of being' and 'huge and mighty Forms that do not live/Like living men mov'd slowly through my mind' (Wordsworth, 1970, lines 418–26). But these frightening experiences of Nature not only reflect the importance that natural objects played in Wordsworth's life, but also show how such objects can be both the object of awe and terror in one context but also of pleasure in another (Winkler, 1975, p. 161). This can be seen to be similar to Darwin's view of Nature: on the one hand pessimistic, in being destructive (for example, through death, extinction and struggle), but optimistic in being creative (for example through diversity, reproduction and adaptation). But the poem is not just relating the experience of the sublime. It is also relating the fact that the child, like Darwin's experience of fossils, rock formations, 'savages', slaves and slave traders, is also experiencing the past of humanity in the present. The child 'is living through the whole life of the race in little, before he lives the life of his century in large [and] may possibly dimly apprehend something more of truth in certain directions than is visible to the adults around him' (Myers, 1899, p. 135). Following Plato, the soul of Wordsworth's child existed in a superior world to Man's before it entered the body,[39] but, at the same time, it was connected by the all-pervading Spirit to the material world surrounding him:

> The child begins by feeling this material world strange to him. But he sees in it, as it were, what he has been accustomed to see; he discerns in it its kinship with the spiritual world which he dimly remembers [...]. And even when this freshness of insight has passed away, it occasionally happens that sights or sounds of unusual beauty or carrying deep associations – a rainbow, a cuckoo's cry, a sunset of extraordinary splendour – will renew for a while this sense of vision and nearness to the spiritual world – a sense which never loses its reality, though with advancing years its presence grows briefer and more rare. (Myers, 1899, p. 135)

Although *The Excursion* impresses upon the reader that the human mind is not part of Nature but is rather accommodated to it as it is 'capable of generating faith, hope, love under the most austere circumstances, of creating beauty [...], of humanising [...] nature and the science that was developing around it' (Gaull, 1979, p. 40), the Mind could nevertheless, as the boy's mental experience on the lake demonstrates, be seen as part of a Primordial Soul.[40] The memories of the poet go back in time through the boy's mind,

---

39   This notion is also expressed in Wordsworth's poem 'Ode: Intimations of Immortality from Recollections of Early Childhood' (Wordsworth, 2015d, unnumbered page). Completed in 1804, published in 1807.
40   Although for Myers the experience of the Primordial Soul is the experience of the human race 'in little' and 'in large'.

which in turn could be seen to be going back to memories of the Primordial Soul before the existence of Man. These memories could be interpreted as representing both the memory of the individual as well as the collective memory of Mankind and primitive life before that. This grasping of this idealized form of reality, or 'realized idealism', is akin to Goethe's archetypal experiences emanating from seeing the sheep's skull and the 'leaf' in Italy. This might appear as a leap from a Platonic Primordial Soul to Romantic Mind as a product of Nature. But the apparent muddle dissipates if one takes a step back and sees all the threads of Nature as Mind, both past, present and future, as an 'entangled bank' made up of the natural history (or 'science') of Nature overlain by the gossamer threads of poetry and imagination. Here Darwin, Goethe, Humboldt and Wordsworth become intertwined. Memories, like 'rastros', are individual histories of the experience of self and Nature but are also memories of physical memories, as with Darwin's experiences of the petrified trees or fossils, or Darwin's experience of the development of the morality of his own era through the abolition of the slave trade. In a metaphysical and a physical sense, all these mental experiences and moral values have come from beyond the physical veil separating us from the inner secrets of Nature – and this beyond could be regarded as the Primordial Soul, as well as a Kantian *intellectus archetypus* in the present.

Like Wordsworth, both Darwin and Humboldt uncover the past from the present through their aesthetic experiences, but as their later works demonstrate, the 'advancing years' makes the experience of the 'nearness to the spiritual world' of Nature 'more rare'. Poetry itself can help Man in getting closer to this spiritual side of Nature, and this was expressed by John Stuart Mill in his autobiography in which he praised Wordsworth's poems for lifting him out of a state of dejection:

> In them I seemed to draw from a source of inward joy, of sympathetic and imaginative pleasure, which could be shared in by all human beings, which had no connexion with struggle or imperfection, but would be made richer by every improvement in the physical or social condition of mankind. (cited by Myers, 1899, p. 136)

This links in to Darwin's theory that morality comes about through the development of instincts of sympathy between fellow human beings which makes them human.[41] Here this uplifting of the human spirit through poetry contributes to Man's 'sympathetic and imaginative pleasure', thereby helping to improve the 'physical or social condition of mankind'. The sympathy of the emotions joining the subjective to the objective, not just in Nature but also between the poet and his readers, reflects a development and maturity of mind, or as expressed by Wordsworth, 'growth', which is in the full title to *The Prelude*, or *Growth of a Poet's Mind*. This mental development of the Mind reflects the moral nature of Man. In his *OUN*, Darwin refers to Mackintosh's *Dissertation* in which he regards this maturity as social, as mankind's fellow-creatures help each other 'for its own sake'. Mackintosh states 'that man at the period of maturity is a social animal, who delights in the society of his fellow-creatures for its own sake, independently of the help and accommodation which it yields' (Mackintosh, 1837, p. 113).[42] Darwin puts this into his own words in his notes when he says that

> These instincts consist of a feeling of love,<and sympathy> <<or benevolence>> to the object in question [...]. We see in other animals they consist in such active sympathy that the individual forgets itself, & aids & defends & acts for others at its own expense. (Barrett, 2008, p. 619)

But the concept of sympathy is strengthened by the word *ought* which makes acts of sympathy towards one's fellow-beings *moral*:

> [The action of] one man trying to save another in desperation [...] shows that our feeling, that the instinct *ought* to be followed is a consequence of that being part of our nature, & its effects lasting, whilst passions although equally natural leave effects not lasting. (Barrett, 2008, p. 620)

---

41 As already referred to, Wordsworth's 'Love of Nature Leads to Love of Mankind' in *The Prelude* depicts a unity of humanity with a single 'spirit' and a single 'moral sense'.
42 Darwin has this passage marked in his personal copy (Barrett, 2008, p. 619, footnote 42–1).

For Darwin the act of trying to save someone from drowning, even though it is obvious that the action will not save that person, is a deeply ingrained *moral* action reflecting a *moral* instinct. It is an action that *ought* to be done for its own sake regardless of outcome. This brings out the 'entangled bank' of mental struggles that exist in Man's mind as he struggles between following strong short-lived passions and weak instincts, 'hence man must have a feeling, that he *ought* to follow certain lines of conduct, & he must soon *necessarily* learn that it is his interest to follow it even when opposed by some natural passion' (Barrett, 2008, p. 621).

Crucial to Darwin's moral law of what is right and wrong, of what are the strong instincts that tell Man what *ought* to be done, is that 'parents' and 'education' are working to achieve the 'same end', namely the benefit of the 'community'. And equally crucial to Darwin's argument, as in the developing physical world, is that the moral laws develop and change according to the beneficial needs of the community. Just as a butterfly species needs to adapt to its changing environment or to other species, so too must Man adapt his moral instincts to changing circumstances within his community:

> As conditions change, from civilization, education changes, & probably likewise instincts, for the same law effects both.- <such> changes <<in accordance to beneficial tendency>> will most readily affect. [sic] the instincts, for they are in accordance with it. (Barrett, 2008, p. 624)

This clearly links in to the earlier discussion of Wordsworth's interrelationship between the subjective and the objective in Nature, in which the Mind has an impact on Nature and Nature has an impact on Mind. Here Nature creates Man's moral code through the development of Man's instincts, yet at the same time through Man's impact on the world he is changing the Nature that creates his instincts. In another sense, Wordsworth saw his poetry as a vehicle for getting his moral point of view across, and so this was an expression of his moral instinct to help better Man's lot through reflection. His *Lyrical Ballads*, produced in collaboration with Coleridge, could be seen as contributing 'to a revolution in taste which had wide social implications' (Scofield, 2003, p. vii). In his poem 'The Last of the Flock', for example, the narrative 'challenges a view of social relations which relies on the condescending charity of the rich [...] with a view which sees human need [...] bound up with the need for self-respect and independence' (Scofield, 2003, p. viii).

In the *Descent*, Darwin explains the existence of the 'moral sense' in humans as having developed through four stages.[43] First, a set of 'social instincts' must have been developed that were strong enough to bind members of society together (Darwin, 2004, 121). Second, the intellect must have been sufficiently developed in order to recall and distinguish the social instinct from a momentary urge that is only concerned with the self (as discussed earlier) (Darwin, 2004, 121). Here Darwin emphasizes the fact that the mental images pass 'through the brain' rather than the mind indicating that the thoughts are physically located despite the fact they are feelings. Also important is Darwin's reference to different mental states interacting with each other, viz. the past and the present. This 'incessant' movement between the two located in the physical present of the brain is a reminder of Goethe's 'Genetic Method' and, in the case of Darwin, his movement between genealogy (the past) and teleology (the future). Third, language is required for community members to express and communicate their needs (Darwin, 2004, 122). But language is not just communication between individuals, it is also a tool used by the community to express its 'approbation and disapprobation' of its members according to feelings of sympathy, which forms part of the collective 'social instinct'. Although expressed physically through speech and the written word, language is a symbolic representation of a collective Mind, yet at the same time anchored in a collective Brain (i.e. an archetypal Brain common to all Mankind). In Darwin's own case, he uses language to reflect his sympathy towards the inhumanity of slavery on the one hand, and the intricacies of Nature on the other, through his 'anthropomorphic descriptions of nature' (Browne 2003a, pp. 213–16). Fourth, and finally, Man's state of Mind (and structure of Brain) has been fully developed to reflexively consider the needs of others within his community through the development of habit ultimately making it instinctive. Man's human state is therefore that of a moral creature (Darwin, 2004, p. 122). In this sense the instincts which are 'obedient to

---

43 See Richard's summary of these stages at Richards, 2002, p. 549.

the wishes and judgement of the community' can be seen as a form of 'community Mind' but physically based in the community's archetypal Brain. When summarizing Adam Smith's argument that 'we are [...] impelled to relieve the sufferings of another, in order that our own painful feelings may be at the same time relieved' (Darwin, 2004, p. 129), Darwin[44] argues that this does not explain why feelings of sympathy are stronger for loved ones and that this 'community of mind' can be narrowed down even further:

> The explanation may lie in the fact that, with all animals, sympathy is directed solely towards the members of the same community, and therefore towards known, and more or less beloved members, but not to all the individuals of the same species. (Darwin, 2004, p. 130)

This feeling of sympathy is an essential part of being human and is highlighted in Mary Shelley's *Frankenstein* when Frankenstein's monster demands the creation of a companion, for living in isolation is a factor in being inhuman: 'My evil passions will have fled, for I shall meet with sympathy! my [sic] life will flow quietly away, and, in my dying moments, I shall not curse my maker' (Shelley, 2015, 144). Communion with others clearly creates Sympathy: '[...]when I live in communion with an equal, I shall feel the affections of a sensitive being, and become linked to the chain of existence and events, from which I am now excluded' (Shelley, 2015, 145). So living in communion with others is an essential part of being human.

So far Man's 'moral sense' has been traced back to Man's intellect, over time, working collaboratively with other members in a community to create a social instinct through habit in which actions are carried out for the common good. The argument presented is that there is a Romantic aspect to Darwin's 'moral sense' as it has been hewn from Nature by Nature rather than having been imposed on Man externally by a Deity. According to the Romantic interpretation of Darwin, the 'moral sense' has been created by Nature's own laws and archetypes. This does not mean that there is not a Deity, only that a Deity has not directly created this 'moral sense'. Darwin

---

44 Darwin refers to chapter one of Adam Smith's *The Theory of Moral Sentiments* (see Smith, 2006).

makes this clear in the epigraphs he uses on the facing page to his title page (Darwin, 1985, p. 50) where he quotes Francis Bacon's work (1605)[45] warning against using the scriptures as a substitute for science, and Whewell[46] in his 'Bridgewater Treatise'[47] who argues for a Deity who makes general laws, not individual beings (Young, 1985, pp. 12–16).[48]

Further evidence will now be given in support of the argument that Darwin was a Romantic Materialist and that Man's 'moral sense' (and therefore his 'reflective' mind) was created from Nature by Nature, that Mind came from matter. Before Man became a moral being, before his intellect was sufficiently developed for him to be a thinking and reflecting being, he learnt to cooperate like other animals in hunting packs. Gradually over time that physical grouping of cooperation had a physical effect. Individuals needed to communicate and thus their vocal cords were developed, and Darwin gives numerous examples of such communication in the insect, animal, bird and marine life. The highest physical evidence of such development that can be both seen by the naked eye and heard by the ear is the human voice that is used for the purpose of language. The development of the vocal cords, and consequently of language, are an example of how the collective cooperative behaviour of humans has brought about a physical change in their organs:

> The habitual use of articulate language is [...] peculiar to man; but he uses, in common with the lower animals, inarticulate cries to express his meaning, aided by gestures and the movements of the muscles of the face [...]. Our cries of pain, fear, surprise, anger, together with their appropriate actions, and the murmur of a mother to her beloved child, are more expressive than any words. (Darwin, 2004, p. 107)

---

45 In Markby, 1863, p. 81.
46 William Whewell's third 'Bridgewater' treatise *On Astronomy and General Physics* (1834).
47 The 'Bridgewater Treatises' are named after the Eighth Earl of Bridgewater (Francis Egerton) who left a will to pay eight gentlemen a thousand pounds each to write a treatise showing the power of God in Nature (Wyhe, 2014, unnumbered page).
48 Although it could be argued that Darwin takes Whewell out of context as Whewell's main point is to argue for a Divine Power (Browne, 2003b, p. 80).

In the same way as instincts are created, as discussed earlier, so habits create physical change. Language started off as a form of cries expressing 'pain, fear, surprise, anger' together with associated body language, which developed the corresponding muscles to articulate those expressions of feeling. And as those muscles develop so too do the articulations and forms of communication get fine-tuned to make them more effective. Thus a cycle of psycho-physical development is created in which the physical expression of a communication affects the organ expressing it and then in turn the improved organ effects a developmental change in the expression, articulation or communication:

> No philologist now supposes that any language has been deliberately invented; it has been slowly and unconsciously developed by many steps. The sounds uttered by birds offer in several respects the nearest analogy to language, for all the members of the same species utter the same instinctive cries expressive of their emotions; and all the kinds which sing, exert their power instinctively; but the actual song, and even the call notes, are learnt from their parents or foster-parents. (Darwin, 2004, p. 108)

This passage could be interpreted as another example of Darwin demonstrating that development comes from within the laws of Nature itself rather than directly from a Deity since the process is slow coming from many steps, and not from one creation at one moment in time. Importantly here, Darwin allows for variation or diversity. Like birds who have developed the same vocal cords within their species yet can learn songs 'from their parents or foster-parents', Man has the freedom within his species to use his vocal cords, and therefore his language, to express himself individually. Crucial to the development of Man's 'moral sense' is the impact that language has on Man's mental development and therefore his brain. The development of language enables Man to 'carry on long trains of thought':

> As the voice was used more and more, the vocal organs would have been strengthened and perfected through the principle of the inherited effects of use; and this would have reacted on the power of speech. But the relation between the continued use of language and the development of the brain, has no doubt been far more important. The mental powers in some early progenitor of man must have been more highly developed than in any existing ape, before even the most imperfect form of speech could have come into use and advancement of this power would have reacted on the mind itself, by enabling and encouraging it to carry on long trains of thought.

> A complex train of thought can no more be carried on without the aid of words, whether spoken or silent, than a long calculation without the use of figures or algebra. (Darwin, 2004, p. 110)

The vocal organs affect speech, speech affects the vocal organs, but most importantly the overall effect of language is to develop the power of thought and therefore the brain. The developed brain affects the development of language and this in turn affects the development of the brain: 'the continued use of language will have reacted on the brain and produced an inherited effect; and this again will have reacted on the improvement of language' (Darwin, 2004, p. 679). As such powers would have been an advantage to the species, the most developed brains would have been selected and passed on to the next generation due to the law of natural selection.

In the first few chapters of this book, Darwin's reference to 'rudimentary organs' in the *Origin* was discussed and showed that along with fossils they created a link between past and present, revealing the history of species' development. As already referred to earlier, Darwin states that 'organs in a rudimentary condition plainly show that an early progenitor had the organ in a fully developed state' (Darwin, 1985, p. 454). These rudimentary organs are present traces of past states of organs that had a use in the past but no longer have a use in the present, such as Darwin's examples of snakes with a rudimentary pelvis and hind limbs. This argument is taken up in a similar vein when discussing vocal cords. Some could be said to be not fully developed whilst others could be regarded as rudimentary:

> As all the higher mammals possess vocal organs, constructed on the same general plan as ours, and used as a means of communication, it was obviously probable that these same organs would be still further developed if the power of communication had to be improved [...]. The fact of the higher apes not using their vocal organs for speech, no doubt depends on their intelligence not having been sufficiently advanced. The possession by them of organs, which with long-continued practice might have been used for speech, although not thus used, is paralleled by the case of many birds which possess organs fitted for singing, though they never sing. Thus, the nightingale and crow have vocal organs similarly constructed, these being used by the former for diversified song, and by the latter only for croaking. (Darwin, 2004, p. 112)

So for example with apes, their 'vocal organs' have not been fully developed to make speech possible and therefore reflects their intelligence as not having been developed sufficiently to make language important for their survival. As humans are related to apes through our tree of descent, this shows how Man's own vocal organs and intelligence must have been rudimentary, and the more rudimentary in time, the more physical and less mentally reflective our states were. The example of the birds with developed organs for singing but not used for song underlines the nature of Darwin's 'tree of life' as it emphasizes the dead ends reached when species stop developing any further. The rudimentary organs either show where the organs were placed in time when useful to a species and now no longer useful, or they show the development of an organ on the way to being something useful but stopping dead in its tracks (a kind of unfulfilled teleology). The important point to make here is that the further back the tree of descent is traced, the more physical and less mental the organisms become. This shows that Mind (both individual, collective and therefore social) came from physical beginnings.

Darwin also shows that language itself has developed gradually and gives examples of rudiments to support this:

> The frequent presence of rudiments, both in languages and in species, is still more remarkable. The letter *m* in the word *am*, means *I*; so that in the expression *I am*, a superfluous and useless rudiment has been retained. In the spelling also of words, letters often remain as the rudiments of ancient forms of pronunciation. (Darwin, 1985, p. 113)

The gradual development of language shows that there 'is no proof that they owe their origin to a special act of creation' and that there can be no 'insuperable objection to the belief that man has been developed from some lower form' (Darwin, 2004, p. 114). And by 'lower form', as discussed above, this can be taken as a more physical form. The development of languages also supports Darwin's theory of natural selection as 'dominant languages and dialects spread widely, and lead to the gradual extinction of other tongues [...]. The survival or preservation of certain favoured words in the struggle for existence is natural selection' (Darwin, 2004, p. 113).

As the above discussion has shown, Mind developed from humble physical beginnings (Darwin's Romantic materialism), showing that morality developed from the material world. This enabled Darwin to provide evidence to support his belief that all races are related through their common progenitor[49] and that although some are more developed than others, they all have the same potential given the right conditions.[50] This is one of the things that informed his opposition to slavery, referred to throughout this book.

Another important argument that Darwin uses in support of his view that Mind comes from matter, is his theory of sexual selection. This is not just a bolt-on theory added to his theory of natural selection but a development of the theory supporting the view that morality developed from matter. Sexual selection[51] might seem an odd example in support of Darwin's view of morality, but as the following discussion will demonstrate, this is another strand of the same argument.

'Natural selection' and 'sexual selection' both come from the same root cause, namely the assertion, as discussed in previous chapters, that Nature is both creator and product of its own laws. However, although they work together, there is a subtle difference between the two: whereas 'sexual selection depends on the success of certain individuals over others of the same sex, in relation to the propagation of the species', 'natural selection depends on the success of both sexes, at all ages, in relation to the general conditions of life' (Darwin, 2004, p. 684). To understand how sexual selection fits into his theory of Mind, it is useful to first look at how Darwin defines sexual selection in the *Descent*:

> Sexual selection has led to the development of secondary sexual characters. It has been shewn that the largest number of vigorous offspring will be reared from the

---

49   'We thus learn that man is descended from a hairy, tailed quadruped' (Darwin, 2004, p. 678).
50   'The mental powers of the higher animals, which are the same in kind with those of man, though so different in degrees, are capable of advancement [...] through natural selection. The same conclusion may be extended to man' (Darwin, 2004, p. 679).
51   That is, those sexual characters that have been developed which 'depends on the advantage which certain individuals have over others of the same sex and species solely in respect of reproduction' (Darwin, 2004, pp. 241–3).

> pairing of the strongest and best-armed males, victorious in contests over other males, with the most vigorous and best-nourished females, which are the first to breed in the spring. If such females select the more attractive, and at the same time vigorous males, they will rear a larger number of offspring than the retarded females, which must pair with the less vigorous and less attractive males. So it will be if the more vigorous males select the more attractive and at the same time healthy and vigorous females; and this will especially hold good if the male defends the female, and aids in providing food for the young. (Darwin, 2004, pp. 255–6)

For the best and strongest to survive and pass on their strengths to the next generation, males need to be armed to fight off other males in competition for the selection of the strongest females that are best at nourishing their young. To do this they require good weaponry, but this same weaponry can also be used to advantage in holding a female for the purpose of successful reproduction. But the males must also be attractive to hold the female's attention and must be strong enough to keep her and protect her and her young. Therefore 'the females [...] prefer pairing with, the more ornamented males, or those which are the best songsters' as well as being 'more vigorous and lively males'. Such pairs would therefore 'have an advantage over others in rearing offspring' (Darwin, 2004, p. 249).

Over time the purely practical organs for individual survival, such as feathers for flight or vocal cords for a warning, start being selected additionally for the advantage they give for furthering the selection of the best mates for reproduction and nourishment and protection of the young (ultimately the best advantage this gives the species for survival). Upon coming across the male Lucanidae beetle with its great mandibles, Darwin felt that 'they are so conspicuous and so elegantly branched, and as owing to their great length they are not well adapted for pinching, the suspicion has crossed my mind that they may in addition serve as an ornament' (Darwin, 2004, p. 345).

At this stage of sexual selection in the insect and animal world there is not an 'intellect' at work as there is in Mankind but there is an element of *mental processing* in which the animal or insect is having to make judgements, however simple, based on the sense data received. For example, a peahen has to identify the longest, most colourful and best-kept tailfeathers when approached by more than one peacock. As discussed earlier, Mankind has developed tools through the development of his intellect.

For example, being able to make a shelter and clothes has enabled Man to do without a hairy body making it easier for him to move between hot and cold terrains. He can adjust his clothing according to the climate. Having tools to make weapons means that he does not have to waste energy growing horns or developing extra-large jaws. The same applies to ornaments. Man can make himself attractive by making ornaments which also gives him the flexibility of improving them immediately without having to wait for the body to develop them over many years. He can also dye his hair and mark his body with tattoos as Darwin discovered amongst 'savages' on his *Beagle* voyage. Darwin is himself aware of the similarities between the ornamentation of birds in the animal world and the ornamentation in Man's world when he refers to women wearing birds' feathers in their hats. Like the theory of natural selection, sexual selection reveals the chain of development, both physical and mental, from animals to humans. As Desmond and Moore put it:

> [Darwin] was satisfied that pea-hens 'admire [the] peacock's tail, as much as we do' and choose the best, showing his belief that aesthetic appreciation could stretch from birds to humans [...]. Going beyond natural selection, [sexual selection] would explain the peacock's fan, as well as those facial and bodily characteristics that made men and women of each race alluring. (Desmond and Moore, 2004, p. xxv)

It is not surprising therefore that 'women everywhere deck themselves with these plumes' confirming 'the beauty of such ornaments' (Darwin, 2004, p. 115).

Reference has already been made to the mind-body development cycle in which the development of the body can develop the mind and, in turn, the development of the mind can further develop the body. For example, the need to cooperate in Man creates the need to communicate, which in turn creates the need to develop the vocal cords which, in turn, develops language and, in turn, enables thought and the development of the intellect. The same applies to sexual selection:

> He who admits the principle of sexual selection will be led to the remarkable conclusion that the nervous system not only regulates most of the existing functions of the body, but has indirectly influenced the progressive development of various bodily structures and of certain mental qualities. Courage, pugnacity, perseverance, strength

and size of body, weapons of all kinds, musical organs, both vocal and instrumental, bright colours and ornamental appendages, have all been indirectly gained by the one sex or the other, through the exertion of choice, the influence of love and jealousy, and the appreciation of the beautiful in sound, colour or form; and these powers of the mind manifestly depend on the development of the brain. (Darwin, 2004, p. 687)

Here Darwin is saying that the development of the nervous system, that is, the brain, and bodily functions, influences the development of the mind which, in turn, through sexual selection, influences the development of the brain, or nervous system. Sexual selection has therefore also been a contributory factor in the development of Mind from humble physical beginnings, creating the 'mental qualities' of 'courage, pugnacity [and] perseverance' as well as the 'mental powers' of 'choice', 'love', 'jealousy' and 'the appreciation of the beautiful in sound, colour or form'.

Like natural selection, sexual selection is also able to reveal the history of physical and mental development linking present-day animals and humans to their primordial past:

> Within the same genus, the two sexes frequently present every gradation from no difference in colour, to so great a difference [that they were long thought to belong to different genuses] [...]. When the sexes nearly resemble each other, this appears due either to the male having transferred his colours to the female, or to the male having retained, or perhaps recovered, the primordial colours of the group. (Darwin, 2004, pp. 355–6)

These 'primordial' traces, like the rudimentary organs, show the historical processes of change linking present forms to past origins. The important point that Darwin is making here is that the past history of these forms is actually visible in the present (that is, at the time he was writing, in the Victorian era) – in a sense the forms are past-present and present-past as they encompass their whole history through gradations of great difference to no difference. The reality of this history is also a reflection of the mental aspect of Nature, since Man's scientific interpretation of Nature through Darwin and others organizes it historically as well as scientifically: through the blurring of differences, 'the primordial colours of the group' shine through with the help of the scientist's imaginative intellect. For Darwin, whether physical or mental, 'the causes have generally been the same which have determined the brilliant colouring of the males alone of some species, and of both sexes of other species' (Darwin, 2004, pp. 355–6). In other words, there is one primordial cause for the outward differences of the two sexes. Darwin gives an example of the Orange Tip butterfly that

> probably shew us the primordial colours of the parent-species of the genus; for both sexes of four or five widely-distributed species are coloured in nearly the same manner [...]. We may here infer that it is the males [...] which have departed from the usual type of the genus. (Darwin, 2004, p. 358)

This is an important argument against Wallace who believed that female lepidoptera adopted their dull colours for the sake of protection.[52] Darwin, on the other hand, believed that the females' colour was the constant and the males deviated from the female colours in order to attract them. In other words, their colours 'have been chiefly modified through sexual selection' which accounts for why 'the females of allied species generally resemble one another so much more closely than do the males' (Darwin, 2004, p. 367). When both females and males are brightly coloured, either imitating species that protect themselves with such markings as a warning that they are unpalatable or imitating their surroundings, Darwin argues that the males have passed on such markings to the females thereby making them deviate from their primordial colours (Darwin, 2004, p. 374).

Darwin uses the example of the male Argus pheasant to support his theory of gradual evolution through simple step changes, from the transition of primary feathers to secondary feathers used as ornamentation to attract females (see Figures 4 to 8). Instead of colours, the male Argus pheasant uses patterns and shapes and these changing forms can be seen between the two types of feathers. This is a living yet fossilized map, frozen in time, showing the tree of descent through Nature's sexual selection.

---

52 Darwin refers to Wallace's paper in the *Westminster Review*, July, 1867, p. 5, in which Wallace states that 'it is only in the tropics, among forests which never lose their foliage, that we find whole groups of birds, whose chief colour is green'. But Darwin counters this by reminding his readers 'that many parrots are ornamented with crimson, blue, and orange tints, which can hardly be protective' (Darwin, 2004, pp. 549–50 and footnote 49, p. 549).

Darwin's Concepts of Morality and Romantic Materialism 157

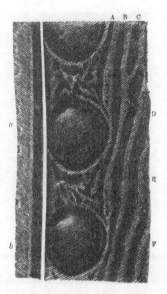

Figure 4: 'Part of secondary wing-feather of Argus pheasant, shewing two perfect ocelli, *a* and *b*. A,B,C,D, &c., are dark stripes running obliquely down, each to an occelus. [Much of the web on both sides, especially to the left of the shaft, has been cut off.]'. Text and drawing from Figure 57 in Darwin, 2004, p. 489.

Figure 5: 'Basal part of the secondary wing-feather [of the Argus pheasant], nearest to the body'. Text and drawing are from Figure 58 in Darwin, 2004, p. 490.

Figure 6: 'Portion of one of the secondary wing-feathers [of the Argus pheasant] near to the body, shewing the so-called elliptic ornaments. The right-hand figure is given merely as a diagram for the sake of the letters of reference. A,B,C,D, &c. Rows of spots running down to and forming the elliptic ornaments. *B*. Lowest spot or mark in row B. *c*. The next succeeding spot or mark in the same row. *d*. Apparently a broken prolongation of the spot *c* in the same row B'.
Text and drawing are from Figure 59 in Darwin, 2004, p. 491.

Figure 7: 'An ocellus in an intermediate condition between the elliptic ornament and the perfect ball-and-socket ocellus'.
Text and drawing are from Figure 60 in Darwin, 2004, p. 492.

Darwin's Concepts of Morality and Romantic Materialism 159

Figure 8: 'Portion near summit of one of the secondary wing-feathers [of the Argus pheasant], bearing perfect ball-and-socket ocelli. *a.* Ornamented upper part. *b.* Uppermost, imperfect ball-and-socket ocellus. (The shading above the white mark on the summit of the ocellus is here a little too dark.). *c.* Perfect occelus'.
Text and drawing are from Figure 61 in Darwin, 2004, p. 494.

Darwin feels that the females possess an 'almost human degree of taste' as it is quite incredible that 'a female bird should be able to appreciate fine shading and exquisite patterns' (Darwin, 2004, p. 449) and that therefore 'the mental powers of birds do not differ fundamentally from ours' (Darwin, 2004, p. 474). The text about the development of these feathers through sexual selection reads like a sacred scroll or tapestry holding all the secrets of the universe; how the pattern itself is not merely a pattern produced by the male to attract the female but is a beautiful pattern of patterns in transition from the practical one of primary function (flight and insulation) to secondary function (ornamentation for the purpose of sexual selection). The beauty is the combined patterns of all the patterns in transition that make up one seamless pattern, but also the pattern that holds all the secrets of transition that reflect gradual evolution. In this respect the tapestry of feathers is sublime in its beauty and akin to a Wordsworth poem:

> Almost every minute detail in the shape and colouring of the ball-and-socket ocelli [of the Argus Pheasant] can be shewn to follow from gradual changes in the elliptic ornaments [...]. A perfect series can be followed, from simple spots to the wonderful ball-and-socket ornaments [...]. The stages in development [...] do not at all necessarily show us the steps passed through by the extinct progenitors of the species; but they probably give us the clue to the actual steps, and they at least prove [...] that a gradation is possible [...]. As the secondary plumes became lengthened through sexual selection, and as the elliptic ornaments increased in diameter, their colours apparently became less bright; and then the ornamentation of the plumes had to be gained by an improvement in the pattern and shading. (Darwin, 2004, pp. 495–6)

In Man this tapestry of mental development can be seen in physical ornamentations which make up the cultures of tribal groups, each having their own idea of beauty:

> In Africa [it is] common practice to raise protuberances by rubbing salt into incisions made in various parts of the body [...]. The natives of the Upper Nile knock out the four front teeth[...]. In the Malay Archipelago the natives file the incisors into points like those of a saw [...]. Further south with the Makalolo, the upper lip is perforated, and a large metal and bamboo ring, called a *pelelé*, is worn in the hole [...]. (Darwin, 2004, pp. 641–3)

Any developments over time within the tribes or cultural groups would also 'modify the character of the tribe' or group and would affect the overall tapestry. But unlike animal breeders, the changes produced would be 'unconscious' as the effect would not be consciously chosen (Darwin, 2004, p. 664). Although in the *Descent* Darwin does not underline the similarity between the moral and cultural ornamentation-sympathies exhibited within communities, the principles binding community members together are the same. Not conforming to the community's moral or aesthetic code brings about disapprobation and isolation preventing the community as a whole from prospering from cooperation. For example, 'when a chief was asked why women wear [*pelelé*], he asked "what kind of a person would she be without the *pelelé*? She would not be a woman at all with a mouth like a man, but no beard"' (Darwin, 2004, pp. 641–3). Missing front teeth, tattoos, 'protuberances' of the skin, and the *pelelé* can all therefore be seen as symbols of social cohesion in the same way as moral behaviours can. Importantly for Darwin, the characteristics of race such as skin and hair colour had been sexually selected as forms of ornamentation, and were not reflections of 'primordial types' created by God for specific regions as proposed by the polygenist Agassiz (he also believed that 'the history of vertebrate life represents the unfolding of a plan of development aimed at man' (Bowler, 1989, p. 128).[53] For Darwin the fact that black people lived in different regions showed that there was no connection between their skin colour and their environment or to an all-powerful God. For Darwin the different skin colour of races was evidence of modification through sexual selection as colour is regarded by Man as 'a highly important element in their beauty':

> We know [...] that the colour of the skin is regarded by the men of all races as a highly important element in their beauty; so that it is a character which would be likely to have been modified through selection [...]. It seems at first sight a monstrous supposition that the jet-blackness of the negro should have been gained through sexual selection; but this view is supported by various analogies, and we know that negroes admire their own colour. (Darwin, 2004, p. 673)[54]

Colour is valued as a form of beauty by Mankind but different races have different standards of beauty and therefore select different colours as their standard:

> If any change [in Man through sexual selection] has thus been effected, it is almost certain that the different races would be differently modified, as each has its own standard of beauty. (Darwin, 2004, p. 653)

As expressed by Desmond and Moore, 'beauty was the main attraction' and therefore 'the only selection necessary was artificial, self-selection according

---

53 Agassiz presented his ideas in the collaborative work *Types of Mankind* [1854] edited by Nott and Gliddon. The polygenists believed that the races could not alter their characteristics in new environments.

54 In arguing by analogy Darwin makes frequent comparisons between Mankind and animals and here blurs the distinction between scientific observer and subjective opinion which could be taken as prejudice: 'The resemblance to a negro in miniature of *Pithecia satanas* with his jet black skin, white rolling eyeballs, and hair parted on the top of the head, is almost ludicrous' (Darwin, 2004, p. 673).

to local preferences' (Desmond and Moore, 2004, pp. xxix–xxx). Darwin extends this argument to the nakedness of skin in Man, particularly women who are generally less hairy than men throughout the world. Darwin puts this down to the fact that the removal of hair enables the colour of the skin to be 'more fully displayed'. The same applies to many birds in which 'it appears as if the head and neck had been divested of feathers through sexual selection, to exhibit the brightly-coloured skin' (Darwin, 2004, p. 669). Darwin explains 'greater hairiness of certain races' as being due to 'partial reversion' to the primordial type. Beards were probably acquired through 'sexual selection as an ornament', and Darwin explains their variability as another example of reversion as 'long lost characters being very apt to vary on reappearance' (Darwin, 2004, pp. 670–2). Whereas the beard was only transmitted to the males, naked skin as an ornament was transmitted to both sexes. The human body through its reversions of hairiness or variations of beards, nakedness, skin colour, hair type or self-imposed ornamentation such as tattoos, knocked out teeth or skin protuberations, all create the tapestry of the history of Mankind in the same way as the feather patterns reflect the history of the Argus pheasant. Like the patterns representing the history of the Argus pheasant, they also provide the key to unlocking the secrets of the development of Mankind, both in Mind and in Body (and therefore in morality and aesthetics) and in so doing provide a sublime experience of Nature through self-awareness (as the history of Mankind runs through the self). This tapestry of the link between Mind and its physical origins can be seen in the common rudimentary organs 'that the embryos of a man, dog, seal, bat, reptile, &c.', share such that at first they can 'hardly be distinguished from each other' and that therefore 'a former progenitor possessed the parts in question in a perfect state' (Darwin, 2004, pp. 42–3). From this Darwin concludes:

> Thus we can understand how it has come to pass that man and all other vertebrate animals have been constructed on the same general model, why they pass through the same early stages of development, and why they retain certain rudiments in common. Consequently we ought frankly to admit their community of descent [...]. (Darwin, 2004, p. 43)

This reflects Darwin's Romantic materialism on two counts. On the one hand it shows the link over time of the development of Man from more 'humble'[55] physical beginnings (that is, the development of mind from body), but it also reveals the extent of Man's mental development through the reflective and imaginative nature of Darwin's own mind in creating this imaginative tapestry through imagery akin to a poem, a painting or a piece of music. Levine regards this imaginative 'humility and honesty [...] almost as a kind of Wordsworthian egoistic sublime, moving from the ordinary against his own original instinct to achieve extraordinary effects' (Levine, 2011, p. 21). Darwin's power of insight can be seen as akin to Wordsworth's in which 'he has shown by the subtle intensity of his own emotion how the contemplation of Nature can be made a revealing agency, like Love or Prayer, – an opening, if indeed there be any opening, into the transcendent world' (Myers, 1899, p. 131), that is, enabling one to 'see into the life of things'.[56]

---

55 But by considering such origins as 'humble' Darwin does not wish to demean their significance or status in the line of descent. In recognising our parentage, Darwin says, we should not 'feel ashamed of it. The most humble organism is something much higher than the inorganic dust under our feet [because such organisms have a] marvellous structure and properties' (Darwin, 2004, p. 193).
56 From Wordsworth's poem 'Tintern Abbey', cited by Myers, 1899, p. 131.

CHAPTER 6

Darwin's Moral and Reflective Nature: Conflicting Values in the Victorian Era

This chapter will examine the question of whether Darwin was a Victorian and whether he could be both a Romantic and a Victorian. Answering the question 'Was Darwin a Romantic?' is an easier question to answer as there are at least Romantic concepts to use as a measuring tool as the previous chapters have demonstrated. However, the question of whether Darwin was a Victorian is a more difficult question to answer. Firstly, Darwin was born in 1809 before Victoria became Queen in 1837, and so clearly he was not brought up as a 'Victorian' and was certainly not a 'Victorian' during the voyage of the *Beagle* between 1831 and 1836, although he lived the majority of his life in the Victorian era. If the question is rephrased to ask 'Was Darwin a Victorian in the sense of holding Victorian values?', this is equally problematical as one could answer 'yes' for some values and 'no' for others. For example, when Darwin published the *Origin* in 1859 this went against the grain of established church values, so in this respect Darwin could be seen as anti-Victorian (yet he was supported by the established 'Victorian' geologist Charles Lyell and the botanist Joseph Hooker). On the other hand he could be regarded as a distinguished Victorian gentleman of independent financial means coming from a respectable line of Darwins and Wedgwoods before him. Yet as a young gentleman in 1841 he was taken into hand by his father Dr Darwin (echoed by his sisters) for not getting his maids to wear caps or for ensuring his manservant, Parslow, had a decent haircut (Burkhardt, 2009b, 296) – outward appearances, even for a naturalist, also reflected position in society.[1] Without respectably dressed

---

[1] The 1861 census for Down House showed that Darwin had twelve servants and the value of his silver was £10,000 at today's prices (Browne, 2003, pp. 457–8).

servants in uniform, Darwin and his wife might be mistaken for 'grocers', and therefore his naturalism might not be taken as seriously.

Darwin's concepts and the Victorian values of the time frequently overlap. For example, Darwin's Malthusian concept of 'struggle' and Townsend's way out of poverty and the workhouse through hard work rather than handouts from the parish have already been referred to. This concept of Self-Help is very much a Victorian value and is exemplified by Samuel Smiles'[2] work first published at the time of Darwin's publication of the *Origin*:

> The spirit of self-help is the root of all genuine growth in the individual [...]. Help from without is often enfeebling in its effects, but help from within invariably invigorates. Whatever is done *for* men or classes, to a certain extent takes away the stimulus and necessity of doing for themselves. (Smiles, 1860, p. 1)

As with Townsend, Smiles believed it was up to the individual to overcome their difficulties and not the government's job, which, in their view, enfeebled the individual. For Smiles, self-help is a motivating factor providing a 'stimulus' in the same way that for Townsend hunger provides a form of stimulation for the poor to do something about it in order to prevent starvation. Although Smiles gives examples of great men such as Josiah Wedgwood, Thomas Carlyle and Robert Stephenson who have achieved great things through self-help, he also states that

> Even the humblest person, who sets before his fellows an example of industry, sobriety, and upright honesty of purpose in life, has a present as well as a future influence upon the well-being of his country. (Smiles, 1860, p. 4)

This Victorian value of self-help is also very Lamarckian. Lamarck argued against immutability believing that individuals could will change through changing habits and that the benefits could be passed on to the next generation. So for example, the short-necked giraffe that stretched its neck more to reach leaves higher up would pass on its longer neck to the next generation, each generation improving the length until the required height was

---

2   Smiles visited Darwin at Down House in 1876 giving him his latest book on self-improvement *Life of a Scotch Naturalist: Thomas Edward* (1876) (Browne, 2003, p. 385).

achieved (Bowler, 1989, p. 86). For the Victorians, this both explained and justified wealth – hard-working men could rise up like Josiah Wedgwood and could, like acquired giraffe necks, pass it on to the next generation, just as the Wedgwoods passed their shooting skills on to Darwin. And those that did not, like Greg's Irish previously referred to, were 'feckless':

> [The] Lamarckian evolution is the perfect metaphor for the self-made rentier class such as the Wedgwoods and Darwins. Owd Wooden Leg Wedgwood[3] [...] made a fortune and was enabled thereby to acquire the houses and lands of a country gentleman. (Wilson, 2003, p. 225)

This view of self-help is also expressed in Chambers' *Vestiges* (1844) in which the Creator created natural laws but not individual creations, thereby leaving Man with the responsibility of his own actions, meaning that the poor could climb out of poverty through their own efforts (Brown, 2003b, p. 21). Before him, the Rev Thomas Chalmers in the first[4] 'Bridgewater Treatise' (1833) and in his *Political Economy* (1832) stated that Man creates his own wretchedness and that 'pauperism' is created by the poor (Young, 1985, p. 53). This was written against Godwin's and Condorcet's view that progress could be achieved through reason and thought enabling Man to achieve harmony and 'perfectibility' which could be passed on to the next generation (Young, 1985, pp. 25–6; p. 39). Wallace's experience encapsulates the alternative Malthusian view of the poor being regarded as irresponsible – his brother William died as a result of poor living conditions although Wallace, despite his struggles, was able to support himself[5] (Browne, 2003b, p. 32). Darwin, on the other hand, through his patriarchal[6] power base of networks within the scientific community, was able to present his joint

---

3 Nickname given to Josiah Wedgwood because of his wooden leg.
4 The first 'Bridgewater Treatise' by Chalmers was *On the Power, Wisdom and Goodness of God as Manifested in the Adaptation of External Nature to the Moral and Intellectual Constitution of Man.*
5 Although he had to collect specimens for the Darwins of the world rather than create an income from his own research.
6 Darwin's patriarchal view of the world is reflected in his *Autobiography* in which he refers to his father and grandfather but misses out his female ancestry (Browne, 2003, p. 427, referring to Broughton, 1999 – page number not cited).

paper[7] with Wallace to the Linnean Society in 1858 without Wallace's knowledge, with his own name at the front, without any serious discussion and without any real effort as it was all done for him in his absence (Browne, 2003b, pp. 18–25; pp. 32–9).

The Victorian era could be summed up as representing a belief in progress and improvement seen in manufacturing, commerce, science, art, technology and social relations, as well as political reform (for example, the 1832 Reform Act extending suffrage to property owners, and the 1867 Reform Act extending suffrage to some members of the working class).[8] The Romantic stretch of the imagination pushing innovation to the limits could be seen as the art of making the impossible become the possible. Apart from the industrial feat of building the railways, this could be observed in anything from the Lipton shop mirrors making customers appear skinny upon entering and well fed upon leaving (Sweet, 2001, p. 53), to Jean François Gravelet's – Blondin's – daring tightrope walks 1,300 feet across Niagara Falls in 1859 (the year of the *Origin*) in which he pushed a wheelbarrow, cooked an omelette and carried his manager on his shoulders (Sweet, 2001, pp. 7–20). With Darwin, this imagination of the impossible could be seen in his experiments in transportation when he hung ducks' feet up to check for any snails (Browne, 2003a, p. 519). This belief in progress and stretching the limits was further expressed through the work of missionaries overseas and charities at home to help the poor improve themselves (Browne, 2003a, p. 248). This view was strengthened by Captain Fitzroy's Fuegians (Jemmy, York and Fuegia) returned to South America on the *Beagle* – their stay in England had enabled them to become virtually fluent in English and this was seen as proof that one could develop from being a 'savage' into an advanced stage of Man (Browne, 2003a, p. 248). Darwin could see the similarities between the Anglicized Fuegians and wild savages and could see a common humanity, which, through evolutionary improvement, stretched right up to his own family, including his sisters (Browne, 2003a, p. 249). Yet his squire-like snobbishness can be seen during his time

---

7   Darwin's paper was sent along with a letter written to Hooker in 1844 to prove that he had the theory before Wallace (Burkhardt, 2009g, p. 507–11, Appendix III).
8   See Wilson, 2003, p. 39 and p. 386.

in South America when he expects Spanish and Portuguese landowners to show off their wealth (Browne, 2003a, p. 262).

So there is a tension and contradiction between Darwin's beliefs as an upper-class gentleman according to his privileged upbringing and position in society, and his more radical belief of transmutation (or evolution) for the time. This can be seen in his attitude towards women, which is also a reflection of the beliefs of the era, as exemplified by the narrative poem *The Angel in the House* by Coventry Patmore[9] (this attitude can be seen to be reflected in the account of sexual selection discussed earlier). Darwin was, for example, condescending towards a Spanish woman dressed for riding like a man (Browne, 2003a, p. 263). Many years later in the *Descent*, Darwin's naturalist 'evidence' mirrors the beliefs of the era, namely that sexual selection has resulted in Man's superiority to women (through Man's competitiveness): 'Man is more courageous, pugnacious and energetic than woman, and has a more inventive genius. His brain is absolutely larger' (Darwin, 2004, p. 622); 'Woman seems to differ from man in mental disposition, chiefly in her greater tenderness and less selfishness' whereas 'man is the rival of other men; he delights in competition, and this leads to ambition which passes too easily into selfishness' (Darwin, 2004, p. 629). 'Imagination and reason' add to man's success and have also been developed through the 'contest of rival males' and 'thus man has ultimately become superior to woman' (Darwin, 2004, p. 631). Although Darwin himself demonstrates a form of 'genius' and 'imagination' in his works, his writing is a mix of the era's beliefs, his own beliefs and the results of scientific research. For example, the reference to Man's brain being larger could be attributed to the prevailing theory of phrenology,[10] or Darwin's view 'that both sexes ought to refrain from marriage if in any marked degree inferior in body or mind' could be seen to support his cousin Francis Galton's

---

9 Coventry Patmore's narrative poem *The Angel in the House* defines the ideal woman as one who pleases Man by being devoted and submissive (Patmore, 2012, p. 68). This ideal was originally expressed by the middle classes but was soon spread further by Queen Victoria's devotion to her husband Albert (Brooklyn, 2011, unnumbered page). A similar ideal can be detected in Emma's devotion to her husband Charles.
10 Already discussed earlier.

theory of eugenics.[11] However, although Darwin recognizes the existing inequalities between the sexes at the time of writing the *Descent*, he does not see them as social inequalities; he sees them as physical and mental inequalities that can only be changed through natural and sexual selection:

> In order that woman should reach the same standard as man, she ought, when nearly adult, to be trained to energy and perseverance, and to have her reason and imagination exercised to the highest point; and then she would probably transmit these qualities chiefly to her adult daughters. All women, however, could not be thus raised, unless during many generations those who excelled in the above robust virtues were married, and produced offspring in larger numbers than other women [...]. Although men do not now fight for their wives, they generally undergo a severe struggle in order to maintain themselves and their families; and this will tend to keep up or even increase their mental powers, and, as a consequence, the present inequality between the sexes. (Darwin, 2004, p. 631)

As the above illustrates, for Darwin the inequalities can be *mitigated* through training in 'energy and perseverance', but this will still not bring women up to the same level of genius and imagination as men. According to Darwin, this can only be achieved over many generations by increasing the number of women who have achieved a higher level of intellect and imagination (Darwin does not elaborate on this but one could imagine some kind of breeding programme if it were not to happen naturally). But Darwin does not see a way out for women to break away from man's dominance and just sees it as totally natural (as perhaps reflected by women's exclusion from suffrage in the 1832, 1867 and 1884 Reform Acts). For Darwin, man naturally protects his family in the face of struggle and this struggle increases his 'mental powers' which means that man will be forever in control. These roles are reflected in Darwin and Emma's relationship, although ironically Darwin's ill health prevented him from being the physically strong provider –

---

11   During the Victorian era, eugenics and social Darwinism fitted in with the attitudes of the time, for example that filth and disease were brought about by the 'immoral' poor. However, this view was mainly held by the upper classes who viewed themselves as superior and used this view to justify their discrimination (Rogers, K. 2009, unnumbered page).

he provided and protected Emma financially through a combination of support from his father and Emma's dowry. Physically and emotionally Emma was the provider as she nursed her husband through illness that dogged his whole life (supporting Darwin's view that woman had 'greater tenderness and less selfishness' than man). For Darwin and Emma, their 'roles were cast from the start. One was a perpetual patient, the other a devoted nurse' (Browne, 2003a, p. 429). These roles are brought out in their letters. For example in a letter to her Aunt Jessie Allen, she states that 'he always tells me just how he feels and never wants to be alone, but continues just as warmly affectionate as ever, so that I feel I am a comfort to him' (cited by Browne, 2003a, p. 430), and in a letter to Emma on 17 November 1848, Darwin writes, 'My own dear wife, I cannot possibly say how beyound [sic] all value your sympathy & affection is to me. – I often fear I must wear you with my unwellnesses & complaints. Your poor old Husband' (Burkhardt, 2009d, p. 183).

Nevertheless, the women in Darwin's household and the women in Victorian Britain can be seen to be in a stage of transition, as part of a social evolutionary process. Women did not have total equality with men, but as industrialization developed, so the demands for equality increased, as did the demands for better working conditions for workers in general. Before the accession of Queen Victoria, women were regarded as 'non-people', having the same status as slaves. Slowly attitudes changed, backed up by legislation such as the Infants and Child Custody Act (1839) giving women access to their children, the Divorce Bill (1855) enabling women to repossess their property, and The Matrimonial Causes Act (1857) allowing women to get divorced from cruel husbands and get maintenance (Wilson, pp. 58–9; Sweet, 2001, pp. 181–2). Although poor women forced into prostitution suffered forcible examinations under The Contagious Diseases Acts (1864), whereas men were not, infants and children were starting to get more legal protection, for example from the Infant Life Protection Bill (1871) after a public outcry over the starving to death of sickly children by paid 'baby farmers'. Darwin could also be seen as controlling his children when using them to assist in his experiments on the one hand, but easing his patriarchal grip on the other in giving them more freedom and time for play (Burkhardt, 2009f, p. XVIII; Sweet, 2001, pp. 176–7). John Ruskin can also be seen to represent this transitional period. In his 'Of Queens Gardens' essay published in *Sesame and Lillies* (1865), he describes the doctrine of 'separate spheres' of man and woman – man being 'creative' and woman being 'domestic'. However, he was also an advocate of progressive education for girls to include such subjects as science, history and mathematics – subjects which had hitherto not been available to them (Sweet, 2001, p. 178).[12] Women were also involved in stylistic editing of books, not just proof-reading.[13] Charles Dickens raised public awareness of the plight of the poor, in particular children, and Charles Kingsley's serialization of the *Water Babies* (1862–1863) satirizes selfish capitalists (the 'Doasyoulikes') who exploit child labour in the form of chimney sweeps, showing that evolution can be regressive[14] (Wilson, 2003, pp. 295–304). This was a period of transition in values and so it would be unfair to expect Darwin to share all the values we would regard as moral and just today. Darwin, for example, went along with the Victorian faith in 'Water Treatments' for the treatment of his probable stress-induced ailments (Browne, 2003a, pp. 492–6), perhaps strengthened by his belief that body and mind were one. This reflects the shift in attitudes towards the concept of Mind and therefore of morality towards the end of the Victorian era. Morality was gradually being acknowledged as manmade (that is, a mental construct from Mind) and therefore independent of God.[15] As the Mind was seen as coming from the brain, the laws of the mind were gradually becoming recognized as a science, viz. psychology (Young, 1985, pp. 57–8; p. 65; pp 71–3).

To understand Darwin's Victorian mind-set, it is important to understand the views expressed in Robert Chambers' anonymously published

---

12 Although Darwin did not arrange a progressive education for his daughters, his sister-in-law, Fanny Wedgwood, supported girls' education (Browne, 2003a, pp. 534–6).
13 Emma Darwin's friend, Georgina Tollet, had considerable input in the style of the *Origin* (Browne, 2003b, p. 76).
14 But Kingsley, like others already referred to, believed that some races were inferior to others and were more akin to animals than humans. He regarded the Irish, for example, as 'white chimpanzees' (cited by Wohl, 1990).
15 In a letter to Asa Gray, Darwin says 'I cannot admit that Man's rudimentary mammae [...] were designed' (cited by Young, 1985, p. 109). See also Young, 1985, pp. 104–5. This is another example of Darwin's belief that such useless rudiments demonstrate that they were not directly designed by a Deity with a purpose in mind. They are merely remnants of Man's developmental process.

Vestiges (1844), as it both summarized the naturalists' ideas at the time as well as the mood of the general public, and that included Darwin despite his criticism of the book. Chambers saw the brain as the 'expansion' of the nervous system ('branching lines'), a continuum of matter from the mollusca and crustacean. He believed that matter formed different grades of 'consciousness of thought', from animals to humans, enabling independent existence (but unlike Darwin, 'designed, formed, and sustained by Almighty Wisdom' (Chambers, 2010, pp. 107–8.) He believed that the difference between animals' and Man's mind was only one of degree, as animals also show signs of 'affection, jealousy, envy' (Chambers, 2010, pp. 108–9). This degree of difference, between animal and Man, was seen to follow a linear advance of development following the 'law of parallelism' in which the growth of a human embryo was thought to pass through the stages of a natural hierarchy from fish to mammal (Bowler, 1989, pp. 126–9). For Chambers, the Almighty controlled the physical laws that made matter possible, and the mind came from the brain which was physical. So the mind was independent of the Almighty yet subject to Divine laws (Chambers, 2010, pp. 105–7). Chambers referred to one part of the brain that is devoted to the sentiments of 'self-esteem [and] love of approbation' (Chambers, 2010, p. 110) which is similar to Darwin's concept of 'sympathy'. Chambers also believed in gradations of mind that were reflections of the stages of development of Man. He believed that negative characteristics such as 'rivalry and jealousy' would disappear as Man became more civilized: 'as civilization advances, reason acquires a greater ascendancy' (Chambers, 2010, pp. 111–20). Moral advancement, for example the elimination of criminality, depended on the influence of members of the community, again reflecting Darwin's concept of 'sympathy' (Chambers, 2010, pp. 115–16). Despite the many similarities to his own views, Darwin attacked the work for not being scholarly enough.[16] Darwin's *Origin* was based on more solid factual information and avoided discussions on the beginning of the earth (Browne, 20013b, p. 61), yet when

---

16 The 'geology strikes me as bad, & his zoology far worse' (cited by Desmond and Moore, 2009, p. 320).

Vestiges was published it was hugely popular amongst the general public and became an obsession (Wilson, 2003, p. 95).[17]

Despite its alleged scholarly shortcomings, there are many similar Romantic expressions of the imagination to Darwin's. When Chambers describes the crystallization of silver and mercury in nitric acid, he compares it to a tree with leaves and roots (Chambers, 2010, p. 53). This is akin to his 'branching lines' in the nervous system and to Darwin's archetypal 'tree', both (through their 'branching') expressing a link to 'vestiges' or 'rudimentary' beginnings ('rudimentary' being a common Darwinian term) and a mind-matter continuum. Like Whewell, Chambers believed natural laws were branches of a 'comprehensive law [which cannot be separated] from Deity itself', and that the Creator did not directly create individual species, as their coming into being was separate from the laws which made this possible. He also believed that moral laws were separate from the laws governing inanimate matter which accounted for disease and disasters (Chambers, 2010, pp. 116–22). Morality was therefore part of Man's will and not God's. Despite elements of Romantic materialism, Chambers reflected the racism of his time, writing that the 'coarse features' of the 'negro race' become a less 'meaner form' if their living conditions are improved. Although he believed that change and development were possible, he believed that the Caucasian was better developed than the 'Negro' and that the Mongolians were regressive and so 'degenerate' (Chambers, 2010, pp. 69–70; pp. 91–102). These views are similar to the contemporary view of the Irish, held by Darwin, already discussed.

The publication of *Vestiges* helped smooth the way for the publication of the *Origin* without this latter having to be published anonymously, although Darwin waited twenty years until he felt confident to get it published. *Vestiges* refers to the everyday world of domesticity, and its similarity in structure to a Walter Scott *Waverley* historical novel made it 'appropriate for a Victorian readership'. What might have once been too dangerous to be discussed in conversation could now be debated in public, 'hence the

---

17   It was read by Queen Victoria, Lincoln, Wallace, Tennyson and George Eliot (Wilson, 2003, p. 95).

wide-spread acknowledgement among contemporary readers that *Vestiges* read like a novel' (Secord, 2000, p. 90). Like Darwin's 'web of affinities' and Humboldt's one reality Nature, *Vestiges*, according to Secord, has 'an underlying vision of history as the product of minute particulars' (Secord, 2000, p. 91). According to *Wade's London Review* in 1845, the author could 'create [this world of particulars] with a dash of his wizard pen ... animalize the dull lump of inorganic matter – and spiritualize, like another Frankenstein, the animal to which his fancy had given birth' (Secord, 2000, p. 41). This Frankensteinian lump of histories can be viewed as reflecting the new industrialization of literary production with a mass readership, of mind and matter intertwined, like one of Darwin's creepers, into affordable popular editions available to the middle classes (Secord, 2000, p. 307). Books like *Vestiges* were not just imparting the latest scientific knowledge to their readers, but were actively engaging them in the discovery of their origins and heritage, both as humans on planet Earth and as part of the cosmos, with or without the helping hand of a Divine power. The narrative is one of exploration and discovery, shared between the narrator and the reader. This sharing of the narrative creates a 'universality' of voice cutting through the divisions of class, age or gender within the industrial society of the Victorian era demonstrating that 'an understanding of nature is accessible to all' (Secord, 2000, pp. 98–9). According to Secord, an example of this sharing can be seen in the use of 'we', in the opening paragraph of page one (Secord, 2000, p. 99):

> It is familiar that the earth which we inhabit is a globe [...] and if we take as the uttermost bounds of this system the orbit of Uranus [...], we shall find that it occupies a portion of space not less than three thousand six hundred millions of miles in extent. (Chambers, 2010, p. 1)

The use of 'It is familiar' also reflects the narrator's understanding of his readership's knowledge, without being condescending (Secord, 2000, p. 100). The story of Nature is 'narrated within a domestic scene', covering such events as birth, childhood, the family and the home, and then extending these domestic images to the solar system with, for example, a description of the planets as 'children of the sun' (Secord, 2000, pp. 101–2). In the same way as adults become interested in their family history today, looking through parish records and old photos, natural history records in the 1840s could shed light on where families and their communities had developed from. The domestic familiarity of the narrative is a subtle way of allaying the fears of the reader by describing the birth of a new species as being no more fearful than the birth of a child (Secord, 2000, p. 108). Similar to Darwin's 'rastro', the term 'vestiges' refers to the Latin *vestigium* meaning footprint, suggesting traces or fragments. Just like Darwin's Indians having to interpret the 'footprints' of other Indians' horses, so too with the readers of *Vestiges* who have to assemble and interpret the fragments of the narrative, and in so doing are indirectly witnessing progress (Secord, 2000, p. 104). This was a time of change and the 'vestiges' of industrialization were also revealing improvements in social and material wellbeing, but not for everyone and not equally.

With the development of the manufacturing industries in the towns and the movement of workers away from agriculture in the rural areas, the lot of the poor was only made worse by the Poor Laws and the Workhouses. Darwin had read Malthus' *Essay on Population* in 1838, and along with the publication of Carlyle's[18] *French Revolution* in 1837, these works contributed to Darwin's concerns for the struggles within mankind, whether this was through overpopulation, revolution or riots (Wilson, 2002, pp. 11–18). The development of capitalism along with an increase in wealth and poverty was also bringing about discontent and demands for better working conditions, along with demands for male suffrage through the People's Charter in 1838. The Depression of 1837–1844 brought the conditions in the workhouses to a head in 1845, with the Andover Workhouse Scandal[19] finally leading to the abolition of the Poor Law commission in 1847. Darwin would

---

18   Carlyle visited Darwin at Down House in 1875 leading to his reluctant acceptance of the theory of evolution, resulting in his comment to Huxley that 'If my progenitor was an ape I will thank you, Mr Huxley, to be polite enough not to mention it [as evolution was] rather a humiliating discovery and the less said about it the better' (Browne, 2003, p. 385).
19   'Colin McDougal, the workhouse supervisor, [...] regularly thrashed children as young as three for messing their beds and he kept his paupers on such short rations that some survived by eating candles' (Wilson, 2002, p. 32).

have been aware of these conditions, particularly as he travelled by train and would have seen more of the poor's lot through his carriage windows (Wilson, 2002, pp. 28–33). As already discussed, the general view at the time was that the poor could help themselves in the same way as 'savages', with encouragement from the missionaries, could improve themselves. Yet in thinking about development and advancement, Darwin did not seem to consider his own privileged position. He assumed that he got to where he got to merely through hard work without reflecting on what underpinned his position. As Browne puts it, 'Darwinism was made by Darwin *and* Victorian society' (Browne, 2003a, p. xi). The poor did not have the same resources or networks that Darwin had to advance themselves. It was not merely a question of hard work:

> Because Darwin believed in the Victorian ethos of character – in the inbuilt advantages of mind – and unconsciously endorsed the cult of great men and public heroes that was so much a part of nineteenth century life, he did not – could not – see that figures like himself were the product of a complex interweaving of personality and opportunity with the movements of the times. Scientific ideas and scientific fame did not come automatically to people who worked hard and collected insects, as Darwin seems to have half hoped they would. A love of natural history could not, on its own, take a governess or a mill-worker to the top of the nineteenth-century intellectual tree. (Browne, 2003a, pp. xi–xii)

This is the paradox. Darwin 'did not' and 'could not' see that he could be 'the product of a complex interweaving of personality and opportunity', yet for today's objective observer looking back in time (which is what Darwin was doing for most of his life), Darwin's own 'personality and opportunity with the movements of the times' can be seen to be akin to the melding and moulding of the 'entangled bank' alongside the rock strata, fossilized past remains and the present diversity of flora and fauna. We the objective observers can see Darwin with his privileged personality and opportunities as part of that very 'entangled bank' that he walked past on his Sand Walk. We can see it, but he could not. The influences that make up the 'web of affinities' in the Victorian era are just as complex as those in Nature. If Darwin were able to describe himself, he would probably agree with Browne's imaginative description of him working away in isolation like one of his own barnacles, with his wife Emma the captain and his children the crew, using his position in society to get everything he wanted through his letters (Browne, 2003, pp. 473–9; pp. 530–2).

This attitude that the poor can easily lift themselves out of poverty through sheer will and hard work is underlined by the British government's attitude towards the Irish during the potato famine of 1845–1850 caused by potato blight. The main issue was the society in which the Irish peasant farmers lived. They relied entirely upon their potato crops in order to survive. Extreme poverty meant that crime figures were high, making the English view the Irish as feckless, dishonest and violent. No consideration was given to the fact that this could be due to their lack of educational or economic advantage or oppression of the mainly Catholic poor by the wealthy Protestants (Wilson, 2003, p. 78). Despite some schemes to import cheap maize from America by the Prime Minister Robert Peel, there was no real effort to take their plight seriously. Prosperous farmers still prosecuted 'starving labourers caught stealing food from their fields' (Wilson, 2003, p. 79). About 1.1 million were thought to have died by starvation between 1845 and 1850 (Wilson, 2003, p. 81). Weakness in the Darwinian jungle was seen as akin to sin. This was underlined by the view of the *Manchester Guardian* which compared the Irish to the English. Unlike the Irish, the English were not starving because 'they bring up their children in habits of frugality, which qualify them for earning their own living, and then send them forth into the world to look for employment' (cited by Wilson, 2003, p. 82). This view of fecklessness can be found in an article by William Greg published in *Fraser's Magazine* in 1868.[20] Greg was a former student friend of Darwin's and Darwin quotes him in support of his view (no doubt held by many other Victorian readers of the *Manchester Guardian* at the time) that the Irish were responsible for their own misfortunes:

> Thus the reckless, degraded, and often vicious members of society, tend to increase at a quicker rate than the provident and generally virtuous members. Or as Mr Greg puts the case: 'The careless, squalid, unaspiring Irishman multiplies like rabbits: the frugal, foreseeing, self-respecting, ambitious Scot, stern in his morality, spiritual in his faith, sagacious and disciplined in his intelligence, passes his best years in struggle

---

20 See footnote 19, in Darwin, 2004, p. 163.

and in celibacy, marries late, and leaves few behind him [...]. In the eternal "struggle for existence", it would be the inferior and *less* favoured race that had prevailed – and prevailed by virtue not of its good qualities but of its faults'. (Darwin, 2004, p. 164)

This view can be seen as even stronger when read alongside Darwin's comments that the 'careful and frugal' are in effect a 'superior class', which he would like to see encouraged rather than providing help and support to the less advantaged. According to this view, it is all down to free will: the poor can dig themselves out of their own hole, which harks back to Townsend's attitude of laissez-faire, referred to earlier:

> A most important obstacle in civilised [sic] countries to an increase in the number of men of a superior class has been strongly insisted on by Mr Greg and Mr Galton,[21] namely, the fact that the very poor and reckless, who are often degraded by vice, almost invariably marry early, whilst the careful and frugal, who are generally otherwise virtuous, marry late in life, so that they may be able to support themselves and their children in comfort. (Darwin, 2004, p. 163)

A similar view of superior classes is mirrored in Darwin's view that it is inevitable that 'the civilized races of man' will ultimately 'exterminate, and replace, the savage races throughout the world' (Darwin, 2004, p. 184) and that

> the break between man and his nearest allies will then be wider, for it will intervene between man in a more civilised [sic] state, as we may hope, even than the Caucasian, and some ape as low as a baboon, instead of as now between the negro or Australian and the gorilla. (Darwin, 2004, p. 184)

---

21 Darwin's cousin Francis Galton argued in his *Hereditary Genius* (1869) that genius runs in families, like the Darwins, and that this could be increased by selective breeding. According to Darwin's son Leonard, Darwin was just as anxious as Galton to 'promote the gradual improvement of our race' (Moore and Desmond, 2004, pp. xlvi–xlvii). However, if Darwin is seen as a 'Romantic' in the Victorian era, this would mean that he should be read *against* eugenics, as Darwin's Romanticism is about Nature improving itself rather than one race or one species gaining superiority or power over all others.

This seems to suggest that Darwin welcomes 'the extermination of savages' as they are less civilized ('as we may hope') and includes 'negroes' and 'Australians' under the term 'savages'. It is important to note that Darwin uses the active verb in 'man will exterminate' rather than the passive 'will have been exterminated' as the result of something else, such as climate. It is also important to note that there is no reference here to the possibility of the savage races being able to become civilized through their own efforts or with the help of missionaries. It is almost as if he is implying that certain races such as the 'negroes' and the 'Australians' are inherently uncivilized and cannot change or improve themselves. This 'racist anti-speciesism' (O'Hear, 1999, p. 134) can also be seen in his preference to be descended from a brave monkey saving the life of a keeper or a brave baboon saving his comrade from a pack of dogs than 'a savage who delights to torture his enemies, offers up bloody sacrifices, practises infanticide without remorse, treats his wives like slaves, knows no decency, and is haunted by the grossest superstitions' (Darwin, 2004, p. 689). O'Hear understands why the Marxists, therefore, saw it as 'no coincidence that [Bishop] Wilberforce [in his attack on Darwin][22] was a Wilberforce, the son of William Wilberforce,

---

22   The 1860 Oxford evolution debate took place at the Oxford University Museum in Oxford, England, on 30 June 1860, seven months after the publication of Charles Darwin's *On the Origin of Species*. Those present included Thomas Huxley, Bishop Samuel Wilberforce, Benjamin Brodie, Joseph Hooker and Robert FitzRoy. The debate is famously remembered today for Huxley's reply to Wilberforce's alleged question as to whether it was through his grandfather or his grandmother that he claimed his descent from a monkey: 'If then, said I, the question is put to me would I rather have a miserable ape for a grandfather or a man highly endowed by nature and possessed of great means & influence & yet who employs these faculties & that influence for the mere purpose of introducing ridicule into a grave scientific discussion I unhesitatingly affirm my preference for the ape' (Desmond and Moore, 2009, p. 497). However, it is important to point out that the recollections of the Huxley-Wilberforce debate are just as famously unreliable. Huxley's alleged reply to Wilberforce is based on his reminiscence written down a couple of decades after the event. There are no other records of the exchange. They no doubt clashed, but it is impossible to state who said what to whom and who witnessed the event other than Huxley.

Darwin's Moral and Reflective Nature: Conflicting Values in the Victorian Era 181

the emancipator of the slaves. Samuel Wilberforce was concerned that Darwinism might give comfort to racial and other supremacists' (O'Hear, 1999, p. 134). But despite this 'racist anti-speciesism', O'Hear does not believe there is any 'reason to suppose that Darwin would have countenanced active genocide, but it is hard to see how the severe struggle he advocates as a means of raising the stock could flourish among a people possessed of universal sympathy and benevolence (O'Hear, 1999, p. 135). When considering Darwin's view of the mutability of species and how species adapt to their environment, this view of fixed classes, either superior or inferior, with pre-set wills of the ability or inability to change their lots, does seem at odds with his theory of natural selection.

However, these views of Darwin's are understandable within the context of the values held by Victorians at the time. The ruthless British reprisals against the Indian mutiny of 1857 could be seen as a Christian civilization putting down an inferior 'barbaric' people (Wilson, 2003, pp. 202–22). This view of the British as a superior race is echoed by prominent literary figures of the time, such as Anthony Trollope who regarded black workers as lazy, Alfred Henty who regarded 'Negroes' as inferior, Jamaica's Governor Edward Eyre who regarded black people as indolent, and Tennyson who called them 'niggers' (Wilson, 2003, pp. 258–72). Rider Haggard's *King Solomon's Mines* (1885) emphasizes cultural superiority in Africa, and this reflects the view of imperialists such as Cecil Rhodes who was encouraged by the British government to create the territories of Rhodesia and the dominion of South Africa (Wilson, 2003, pp. 601–6). This Victorian attitude of treating other races as inferior could be seen in battle and continued after the death of Darwin, in 1900–1901, with Kitchener's burning of farms and the starving of women and children in concentration camps during the Boer Wars (Wilson, 2003, pp. 612–13).

When considering Darwin's 'reflective' thinking within the context of Victorian values as outlined above, Darwin appears to blur the distinction between two forms. The first is the 'reflection' on objects in the physical world consisting of the processing of sense data in time and space creating the 'I' or ego; and the second is the 'reflection' on the reflection-making process creating emotions or feelings felt by the observer (that is, feelings that amount to the moral compass of the self and community). The first form shows that Man's mind originates from material nature (a theory that Darwin is trying to prove from scientific facts) – this is the '*is*' form of reflection. The second form is Darwin's moral standpoint, the '*ought*' form of 'reflection': on the one hand Man is equal through his common progenitor, and on the other some races are better developed than others and the better developed act as a natural check against the less well developed. But Darwin's reference to Greg suggests that he *might* be sympathetic to those wishing to nudge Nature on a bit through human intervention. Darwin's reasons for not supporting eugenics, that is, coercive selective breeding, could be interpreted as pragmatic, as eugenics could have a counterproductive moral effect on those doing the coercing, in the same way as forcing people to inflict pain on others would have the effect of damaging their moral compass. Yet Darwin can be seen to be sympathetic to non-coercive means such as selecting one's mate carefully – Darwin was very much aware of this in his own case of selecting his cousin Emma Wedgwood as his wife and the possible effect of this on the health of his children. Either way, there is a dangerous tendency for Darwin to blur the distinction between scientific evidence and subjective moral views. A worrying reference in *The Descent* refers to reversion in which a characteristic, like the stripe of the zebra occurring occasionally in horses, is likened to the black sheep of the family – that is the undesirable inherited characteristics of the family that reveal themselves in an individual:

> [...] injurious characters which tend to reappear through reversion, such as blackness in sheep; and with mankind some of the worst dispositions, which occasionally without any assignable cause make their appearance in families, may perhaps be reversions to a savage state, from which we are not removed by very many generations. This view seems indeed recognised [sic] in the common expression that such men are the black sheep of the family. (Darwin, 2004, p. 163)

Although Darwin does not actually say so, he gives the impression that he would happily see certain undesirable characteristics weeded out.[23] If this

---

23 Reading the *Descent* as a whole, it can be seen that for Darwin any coercive means of weeding out undesirable characteristics would be counterproductive as this would reduce the level of 'sympathy' among members of the public doing the coercing.

was Darwin's view it would indeed be a darker side of his Romanticism and could be regarded as an example of a 'black' reversionary streak of human nature, as a form of primitive savagery dressed up in the guise of Victorian respectability.

Darwin's own nature is the result of his privileged background acting as a psycho-socio-cultural constraint, inhibiting insightful reflection of his own nature and society; unable to see that his own privileged class takes advantage of its position over the less fortunate classes; unable to see that such prejudice against the Irish is akin to the same prejudice he deplores against blacks and slaves; and that this class dominance used to make his class stronger and the weaker classes weaker runs in his own veins.

Is Darwin the thinker more Romantic than Darwin the writer, and is Darwin the writer more Victorian than Darwin the thinker? It can only be reiterated that there are only *aspects* of Darwin that can be labelled 'Romantic' or 'Victorian'. In other words, there is not one colour running through Darwin's 'stick of rock' – he is a blend of Romantic and Victorian colours even though on the surface they might appear to be contradictory. For example, as already discussed, the Victorian 'colours' can be reflected in his belief that women are less intellectual than men (yet Darwin regarded men and women as equal in that they come from the same progenitor and, as he has demonstrated, the male butterfly colours come from the original base female colour). The Victorian belief that certain races were inferior such as the Irish, the Sepoys in India and the 'savages' in America is another example (yet Darwin's notion that Nature was able to adapt and improve ran in parallel to the work carried out by the missionaries in America, although imperialist in nature, showing that the Victorians *did* also believe that change and improvement *was* possible. This is also reflected in political reform such as the Reform Acts and social change such as the Factory Acts, which underpinned the improvement of working conditions). At the centre of the mix, where the two colours blur into

---

Such a reduction in this 'sympathy' would therefore reduce the level of morality in that society, therefore making it less civilized or morally advanced than before. 'A just public opinion' is therefore very much a part of the concept of 'sympathy' and morality (Darwin, 2004, p. 682).

one, is the Victorian naturalist on the one hand bringing back crates full of specimens, dead or alive, normally used to fill museums or zoos, and on the other hand a new breed of aesthetic scientist akin to Humboldt who is able to lift his imagination beyond the scientific data to something equally poetic and artistic; in other words, the imaginative ability to see the 'web of affinities'. As demonstrated above, Darwin is a complex mix of the Romantic and the Victorian, both at the conceptual and periodic levels. The conceptual distinction covers Darwin's absorption of ideas from both the historical periods of Romanticism and Victorianism given how he comes of age intellectually on the cusp of the Romantic into the Victorian period.

As the preceding discussion has already demonstrated, there are various influences that pull Darwin in one direction or the other. Darwin's most insightful experiences of pure imagination were obtained when he was able to experience Nature first hand, without the constraining or restraining influences of the values of his era (although it could be argued that all experience is coloured by society's values to an extent). The most imaginative can be seen in the 'Humboldtian' experiences of South America. However, Darwin did have such aesthetic, almost mystical, experiences before he voyaged on the *Beagle*. One of these was while he was a Cambridge student when he was uplifted by his aesthetic reflection of Sebastiano del Piombo's painting *The Raising of Lazarus*[24] (1517–1519)[25] exhibited at the National Gallery, London (see Figure 9). Darwin's friend at university, Whitley,[26] frequently went with him to the National Gallery in London, and in his autobiography (1876) Darwin wrote 'of Sebastiano del Piombo exciting in me a sense of sublimity' (Darwin, 1995, p. 19). Through Whitley he learned

---

24 This miracle of raising Lazarus from the dead attributed to Jesus in the Gospel of St John 11:1–44 demonstrates his divine power over death, one of the enemies of humanity. The name 'Lazarus' comes from the Hebrew, meaning 'God is my help'.
25 Sebastiano's *Raising of Lazarus* was commissioned by the archbishop of Narbonne in France to match Raphael's *Transfiguration* and was in competition for a place in a church in Rome. The painting cast Christ in the role of healer and divine physician to raise Lazarus from the dead four days after his death (National Gallery, 2014).
26 Whitley also came from Shrewsbury school (Browne, 2003a, pp. 105–6).

Darwin's Moral and Reflective Nature: Conflicting Values in the Victorian Era 185

Figure 9: *The Raising of Lazarus* (1517–1519) by Sebastiano del Piombo. The image has been reproduced with the kind permission of the National Gallery, London. © The National Gallery, London.

the art of looking, of deciphering the layers of the painter's technique and the allusion behind the technique.[27] It enabled him to expand his imagination through aesthetic experience and to apply this to his experience of Nature itself. Neither science nor the religious theme of the painting held him back from experiencing the sublime (Browne, 2003a, p. 106). The accompanying text from St John's Gospel also encapsulated an experience of the wonder of Nature through the complex emotions of the polar opposites of faith and disbelief that go with the thoughts of creation and resurrection:

> Sebastiano's text from St John's Gospel went straight to the heart of Christianity, encompassing creation, resurrection, faith, and disbelief: 'I am the resurrection, and the life'. For the impressionable young Darwin, the sense of intense wonder was as readily inspired in an artistic and theological context as in exploring nature. (Browne, 2003a, p. 106)

Darwin's interest in the painting could also be seen as a reflection of his fascination with death and extinction through his interest in fossils.

Yet, despite such subliminal imaginative experiences, both in the art gallery, in the jungles of South America and in his studies in London and Kent, it can be argued that Darwin's Victorian values[28] prevented him from gaining a similar insight into the dehumanising streak of prejudice existing in himself and his own class (as already referred to in the discussion on Victorian values,). If he had gained such an insight, he might have experienced Kurtz's 'The horror! The horror!' (in Conrad's *Heart of Darkness*), when confronted with Man's darkest side of uncontrolled human nature. Perhaps to an extent, Darwin's move to Down House was a way of burying himself not only in his work away from the distractions of London but also of burying himself in isolation away from the Victorian values that constrained him (like a metaphorical *Rhea darwinii* burying its head in

---

27   As already discussed, Darwin's *Origin of Species* can be seen to be a representation of a Phoenix-like rebirth of fossilized remains. This rebirth is mirrored in *The Raising of Lazarus*, both in terms of Lazarus and the materials used to paint him.
28   This term is not used in the sense of the term popularized by Margaret Thatcher's administration in the 1980s regarding liberalism and small government.

his 'Sandwalk').²⁹ Doing his research independently financed by himself meant that he also did not have to kowtow to any academic institutions or the government. So in this sense he could be said to be less Victorian than his peers. Other personal factors also made him less Victorian and less dependent on institutions. The death of his daughter Annie,³⁰ for example, which he never got over, strengthened his disbelief in a traditional God,³¹ as did the untimely deaths of other members of his family and friends around this time.³² But none of these experiences enabled him to step out of himself in order to look into his own mind, to see how he and his class fitted into his own line of 'Descent'.

In conclusion to this chapter, despite Darwin's 'reflective' shortcomings, Darwin's *Origin* and *Descent* could be seen as a written equivalent of *The Raising of Lazarus*, a Phoenix rising up from the dead, a rebirth of a fossilized past made up of data in the present. In the same way as the thousands and thousands of brush strokes make up the picture alongside the layers of canvas, paper and glue, the written history of the past makes up the present through the structure of its 'tree' of descent, its branches and constituent species. In both cases, death rises up into the present as life. Through Darwin's image of the past, the reader gets as close as he can

---

29   The famous Sandwalk is where Darwin would take his exercise at Down House using a pile of flint stones to mark his five circuits making up a half mile. This 'was his spiritual home [...] where he pondered. In this soothing routine, the power of place became preeminent in Darwin's science. It shaped his identity as a thinker' (Browne, 2003, p. 402).

30   Anne Elizabeth "Annie" Darwin 2 March 1841–23 April 1851. In his memorial to her, Darwin wrote: 'We have lost the joy of the Household, and the solace of our old age:- she must have known how we loved her; oh that she could now know how deeply, how tenderly we do still & shall ever love her dear joyous face. Blessings on her. – April 30. 1851' (Burkhardt, 2009e, p. 542).

31   'Annie's cruel death destroyed Charles's tatters of belief in a moral, just universe [...]. Charles now took his stand as an unbeliever' (Desmond and Moore, 2009, p. 387). 'This death was the formal beginning of Darwin's conscious dissociation from believing in the traditional figure of God' (Browne, 2003a, p. 503).

32   For example, the death of his father Dr Robert Darwin on 13 November 1848, the death of his sixteen-month-old son, Charles Waring Darwin on 28 June 1858, and the death of his friend and mentor John Stevens Henslow's wife Harriet in 1857.

to the concept of his own non-existence on the one hand, and close to the continuation of his existence through the human species on the other. Yet even Man's non-existence can be experienced through the idea of the connected existence of simple organisms before that. The flight of the Phoenix, therefore, seems to fly in both the direction of the past and the future. *The Raising of Lazarus* as a physical painting can also be seen as Darwinesque as it is made up of layers of past restorations. Each restoration is made up of its own layers of glues, papers, boards, varnishes and flakes of paint precariously held together, each with their own history (due to blisters, shrinking or bug-infested glues, grime and so on). Again past and present combine and evolve to create a form moving into the future hinged on historical restoration techniques and technologies.[33]

The above discussion has demonstrated that Darwin can be regarded as a Romantic materialist. For Darwin, Mind comes from the origins of matter. The highest forms of Mind are Morals gained through instinct and imagination. Conscience guides moral choice in being able to move freely between past and future actions, in being able to assess sympathies towards fellow Man and feelings of sympathy from fellow Man to oneself. In a similar way, the flight of the imagination, like a bird (or 'castles in the air') is also able to move between the past and the present, but also between subject and object, between the particular and the universal, between the empirical and the scientific, and between the humble and the sublime. In the form of Darwin's work, like Wordsworth's poetry, it can be seen as a 'double consciousness', reflective through a poetic dialogue or 'conversation', like the 'Humboldtian Method' and Goethe's 'Genetic Method'. The metaphysical form of the archetypes helps the Mind to conceptualize Nature and helps the scientist and the poet to get closer to an understanding of it, at its highest form approaching a mystical or sublime experience. Despite Darwin's reflective shortcomings in terms of understanding his own contradictions in terms of 'racist anti-speciesism' (which could be

---

33   For a history of the painting's restoration work, see Dunkerton and Howard, 2009, pp. 26–51.

due to the values of the Victorian era in which he lived),[34] Darwin's highly tuned imagination can be shown to have unlocked the door to some of Nature's most intimate secrets. His imagination along with his writings, (like a Wordsworthian poem), can be seen as a phoenix rising from the primordial past just like the feathers of the Argus pheasant or the restored painting of *The Raising of Lazarus*.

---

34 The metaphor of the 'entangled bank', as discussed in this chapter, can be seen as a way of explaining Darwin's inability to stand outside himself.

CHAPTER 7

The Transmutation of Darwin's Romanticism

This chapter analyses the text of the *Voyage of the Beagle* and Darwin's *Beagle* notebooks in relation to Darwin's Romantic concepts, discussed in earlier chapters, to see how they were developed in the early Darwin, to see how the language was used to express them, and how the form of Darwin's expression changed as he matured in years, as reflected in the *Origin* and the *Descent*.

Darwin's early piecing together of the history of Nature has already been described as similar to a South American Indian following the tracks (or 'rastro') of horses. This reading of Nature creates the narrative of Nature, and throughout Darwin's writing the reader is presented with a 'double movement of prose' in which he is first presented with an impossible-to-believe discovery (for example, shells found at the top of a mountain) creating the experience of wonder, followed by an explanation (for example, the sea beds were upheaved over millions of years to form mountains) creating yet more wonder that such a thing could actually happen (often followed by an exclamation mark). The narrative of Darwin's works reveals not only the development of Man's Mind from simple beginnings, but also the development of Darwin's Mind alongside that of the reader's. The 'rastro' reading of Nature along with the 'double movement of prose' is an experience shared by writer and reader and this chapter will examine these two aspects of Darwin's writing in order to further understand his imaginative and reflective thinking.

Reference has already been made to the influence of Wordsworth, the importance of poetry in Darwin's youth and how in later life he no longer enjoyed it. The following analysis examines the importance poetry played in developing his imagination through the experience of Nature creating a personal poetic experience, and how in turn this was transformed into a more objective poetry of science, capturing both the wonder and enchantment of Nature. This development of a poetic imagination can be seen to reflect a move away from an anthropocentric world in which Man is the centre, to a more anthropomorphic world in which Man and the animal world share common sympathies of consciousness; in which humble creatures such as the common earth worm are given traits that are almost human. Darwin makes the small and insignificant appear big and sublime.

Darwin's analysis of the data that make up the history of Nature is not merely the result of deductive reasoning; it is also a process of poetic imagination in which the impossible is allowed expression (the 'castles in the air' referred to in previous chapters), combined with the ability to focus on detail, the unusual rather than the normal, and to make comparisons between species, geographical regions and periods of time. This is Romantic in the sense that it incorporates the 'oneness' of Nature in which everything is related to everything else, and here, in relation to the 'rastro', includes the relationship between the objective forms of Nature being pieced together by the subjective 'reader' of Nature; that is, both Darwin, reading Nature and writing the narrative of Nature, and the reader of Darwin's works. This is one of the highest forms of Mind that has developed from the simple beginnings of instinct, in turn developed from matter. Darwin's 'rastro' method of reading Nature may also be seen to be Romantic in its ability, through the medium of poetic prose, to paradoxically unite the possible with the impossible, as with uniting Mind and matter, to make the mystical tangible. And these 'rastros', or traces, that point towards natural selection and evolution are not just footprints, but include behaviours and blood stains as well as fossils:

> Homologies, behaviours, fossils, distributions, embryos, and more – these are the footprints. And they point uniquely to one culprit: evolution through natural selection! (Ruse, 2009, p. 142)

This skill of learning to 'see' through paying attention to detail can be seen to have been developed through reading Charles Lyell's *Principles of Geology* whilst on the *Beagle* (Levine, 2011, pp. 44–50). Constructing hypotheses based on observable causes (such as the volcano at Etna in Sicily) was seen as a better way of explaining phenomena than on unobservable catastrophes (Bowler, 1989, p. 135). Lyell revived Hutton's steady-state system which suggested a balance of constructive and destructive forces through time, such as the erosion of land surfaces by the flow of water creating sediment, and the elevation of mountains through earthquakes (Bowler, 1989, p. 136). Darwin completed reading Lyell's third and final volume of the *Principles* in April–June 1834, which helped reinforce Darwin's view that the elevation of Patagonia had been gradual and that 'many successive earthquakes in the Andes' had a 'potential elevating power' (Chancellor, 2009, p. 136). But this understanding of Nature is not just obtained through raw perception, but through a combination of reason and poetic imagination. As Levine puts it, 'raw perception is always tangled with reasoning' and reason creates wonder through poetic insight (through the 'double movement of his prose', moving from wonder to explanation and to wonder again) (Levine, 2011, pp. 52–61). This wonder is not just created by being struck by awe but is created by spotting the unusual and asking questions (Levine, 2011, pp. 53–5). Through this 'double movement of prose', an initially self-effacing Darwin expresses wonder at the unknown and the unknowable, yet then explains natural phenomena through the ordinary, creating the feeling of wonder again that such a thing could be possible. For example, in the *Origin* Darwin expresses his wonder at the instinct of the hive bee to create the hive. He then explains the development of the ordinary bee from working independently to working with other bees to create a hive and how this instinct is passed on (creating the feeling of wonder again). The exquisite structure of the hive bee has been built up through natural selection by 'imperceptible gradations'. Such gradations can be seen in bees today with varying degrees of perfection (Levine, 2011, pp. 156–7). As expressed by Levine, this is the perfect paradox as the individual bees consciously perform the act of making a hive without knowing what they are doing (Levine, 2011, p. 158): 'The bees [...] no more knowing that they swept their spheres at one particular distance from each other, than they know what are the several angles of the hexagonal prisms and of the basal rhombic plates' (Darwin, 1985, p. 256). The paradox can equally apply to Darwin as he is conscious of the details of Nature as it unfolds before him yet is ignorant of the history in which he and everything around him is embedded. His theory is an imaginative leap based on his experience of data creating a history, a 'rastro', of which he is a part but of which he is not fully conscious. Nature produces beautiful and complex structures without being aware of what it is doing: 'Not only instinct but human consciousness itself grows from mindlessness' (Levine, 2011, p. 159).

Levine (2011, p. 61) identifies the same 'rastro' method of tracing history used by Darwin when he refers to the connection Darwin makes between the 'mud, sand and shingle' of the beaches with the 'rattling noise' of the stones in the mountain torrents:[1]

> As often as I have seen beds of mud, sand, and shingle, accumulated to the thickness of many thousand feet, I have felt inclined to exclaim that causes, such as the present rivers and the present beaches, could never have ground down and produced such masses. But, on the other hand, when listening to the rattling noise of these torrents, and calling to mind that whole races of animals have passed away from the face of the earth, and that during this whole period, night and day, these stones have gone rattling onwards in their course, I have thought to myself, can any mountains, any continent, withstand such waste? (Darwin, 2003, p. 316)[2]

Again Darwin uses the 'double movement of prose' in which it seems impossible that the mountain torrents could produce 'mud, sand, and shingle, accumulated to the thickness of many thousand feet'. But then when he contemplates the huge period of time in which 'whole races of animals have passed away from the face of the earth', he realizes that this same mountain torrent *would* have continued bringing down its 'rattling stones' and therefore be able to produce 'such waste'.

Darwin's description[3] of the contrast between the noisy torrents of the mountain and the 'tranquillity' of the beaches below also mirror the parallel image of vibrant living species in struggle on the one hand, and the tranquillity or peace of death and extinction on the other hand. The

---

[1] Whilst in the Passage of the Cordillera on the 19 March 1835.
[2] This edition of the *Voyage* is referred to as the passage does not appear in the Darwin 1989 publication (1839 first edition), which is the main text referred to. This applies to further references to the 2003 publication (1860 third edition).
[3] Described in September 1833 while travelling from Bahia Blanca to Buenos Ayres. This description of the deep past in the present echoes James Hutton's steady-state theory of Uniformitarianism, made more widely known by John Playfair and probably percolated down to Darwin via Lyell (Bowler, 1989, p. 46).

disappearance of the waters has been slow and has not been due to any 'great catastrophe':

> I do not think nature ever made a more solitary, desolate pile of rock [...]. The strange aspect of this mountain is contrasted by the sea-like plain, which not only abuts against its steep sides, but likewise separates the parallel ranges. The uniformity of the colouring gives, also, an extreme quietness to the view; the whitish gray of the quartz rock, and the light brown of the withered grass of the plain, being unrelieved by any brighter tint. From custom, one expects to see in the neighbourhood of a lofty and bold mountain a broken country, strewed over with huge fragments. Here nature shows, that the last movement before the bed of the sea is changed into dry land, may sometimes be one of tranquillity. Under these circumstances, I was curious to observe how far from the parent rock any pebbles could be found. On the shores of Bahia Blanca, and near the settlement, there were some quartz, which certainly must have come from this source: the distance is forty-five miles. (Darwin, 1989, pp. 116–17)

The 'tranquillity' or 'extreme quietness' of the beaches after the noisy torrents have stopped is also expressed by 'the uniformity of colouring' such as the 'whitish gray of the quartz rock' and 'the light brown of the withered grass' (here the 'withered' denotes the ending of the previous life of the terrain), all quiet, toned-down colours rather than a 'brighter tint' which might suggest more violent forces at work (such as the 'noisy torrents'). The quietness of the colours also contrasts with the expectation of a violent terrain normally associated with 'a lofty and bold mountain', such as 'a broken country, strewed over with huge fragments' (Levine, 2011, pp. 62–3). Once again, Darwin is hit by wonder by the opposite of what he expected, by the impossible that becomes a reality. Finally, the enormous change that Darwin sees at the end of the torrent's journey is reflected in the 'distance of forty-five miles' that the quartz has travelled from its source, and although now 'tranquil' hides the rattling noise that once occurred (Levine, 2011, pp. 62–3). Darwin's 'rastro' is not a mere forensic Sherlock Holmes-type descriptive analysis. It is also a multi-layered work of poetic art, both in terms of its poetry and imagination, but also as a kind of visual art that comes through the imaginative description. The reference to the quiet colours, for example, gives the prose a perspective like a painting, making the colours fade away into the distance – and the distance created by this perspective is an exact reflection of the distance between the start of the torrent and its end on the beaches. This is a work of art captured in prose, but when imagined visually can be seen as a landscape painting (with Darwin the observer included in the landscape). This is akin to the Romantic Argus feathers in the *Descent*, representing the same 'rastro' traces of history, with their depth of perspective measuring out the distance of time between past and present through the changing patterns and shades of the ocelli.

When viewing the 'streams of stones' in the present moment that over time have created the tranquil beaches below,[4] Darwin also compares them to his past experience of the after-effects of the earthquake at Concepcion: the 'stupendous mountains have been broken into pieces like so much thin crust' (Darwin, 2003, p. 197). Using the 'rastro' method, Darwin again relates the present to the past but also through the 'double movement of prose' argues that awe-inspiring phenomena that seem impossible to explain can potentially be explained naturalistically rather than using a deity as cause:

> Never did any scene, like these 'streams of stones', so forcibly convey to my mind the idea of a convulsion, of which in historical records we might in vain seek for any counterpart: yet the progress of knowledge will probably some day give a simple explanation of this phenomenon, as it already has of the so long-thought inexplicable transportal of the erratic boulders, which are strewed over the plains of Europe. (Darwin, 2003, p. 197)

Darwin may have had Stonehenge[5] in mind when referring to 'the erratic boulders [...] strewed over the plains of Europe' (Levine, 2011, pp. 66–77), that can be explained yet at the same time imagined as if they are 'erratic boulders' before being assembled by Man. Geologically the term 'erratic boulders' suggests that they have been brought from a distance by glacial action, but for the reader familiar with Stonehenge, the same contrast between 'tranquillity' and 'noise' or violence can be imagined: the majesty of the towering stones against the sky or a sunset, the mystical experience

---

4   Written on the 19 May 1834 whilst visiting the Falkland Islands.
5   In the late 1870s Darwin and Emma visited Stonehenge, and armed with a spade, went looking for worms (Browne, 2003, pp. 447–8).

of their arrangement, all creating a feeling of peace and tranquillity; yet at the same time they are 'erratic boulders' (rather than stones hewn by Man) that have been 'strewed over the plains of Europe' as if by that same earthquake at Concepcion, thus creating the same impression as that of the 'streams of stones' conveying 'the idea of a convulsion'. It is not difficult to imagine Darwin excavating worms from under these stones yet at the same time looking up at the towering stones above him, trembling under their power of the sublime, experiencing a mixture of the tranquillity of their present stillness with the distant rumblings of their movement in getting the stones to their present positions, however that was achieved. Darwin often explained natural phenomena through his imaginative use of analogy and so it is not difficult to imagine Darwin imagining glacial action at Stonehenge even though he is aware that the boulders were moved by Man. Even the impression they create in the mind is a violent one, as expressed in 'so forcibly convey to my mind'. Darwin is not just using the 'rastro' as a forensic tool to analyse his data, but subtly creating a narrative poem as a piece of conceptual art capturing all the shades, colours and textures of his images, all helping to bring the past histories alive in the present.

Darwin also uses the 'rastro' method to propel himself into the past to imagine the way things might have happened as a way of explaining the present. For example, when Darwin sees the shattered and broken rocks in an area where there are no longer any earthquakes appearing, he imagines them being 'hurled into their present position thousands of years ago'. At the summit of Mount Wellington in Van Diemen's Land in Chile,[6] Darwin wonders 'at the force which has upheaved these mountains, and even more so at the countless ages which it must have required, to have broken through, removed, and levelled whole masses of them' (Darwin, 2003, p. 257; Levine, 2011, p. 67). Darwin develops this form of imaginative 'rastro' in the *Origin* as an explanation to help him interpret the traces of previous glacial action (Levine, 2011, p. 97):

> But we shall follow the changes more readily, by supposing a new glacial period to come slowly on, and then pass away, as formerly occurred. As the cold came on, and

6 On the 17 August 1834.

as each more southern zone became fitted for arctic beings and ill-fitted for their former more temperate inhabitants, the latter would be supplanted and arctic productions would take their places. (Darwin, 1985, p. 360)

The 'rastro' method also helps Darwin discover 'the infinite interdependence of organisms on each other, on the weather, on time, on geology' (Levine, 2011, p. 106). This 'interdependence' is another example of Darwin's Romanticism as it reflects the 'oneness' of Nature in which everything is related to everything else, in which everything has an equal part to play, whether this be Man or the humble worm. This form of the 'rastro' is developed further in the *Origin* in which Darwin's understanding of the interdependence of organisms in one area helps him understand them in another area (in a similar way to the 'streams of stones' helping Darwin understand earthquakes better). Darwin illustrates this by comparing Staffordshire with Farnham, Surrey. In Staffordshire, where Scotch fir trees had been enclosed, the insect and bird population had increased but not in the adjoining heathland. In Farnham, Darwin could see the difference between an enclosed area of Scotch fir where there were self-sown trees, and an unenclosed area where seedlings were browsed by cattle and were unable to grow (Levine, 2011, pp. 107–8; Darwin, 1985, pp. 123–4). Staffordshire and Farnham can also be related to Paraguay: flies laying eggs in the navels of cattle reduce their reproduction – the balance is kept by the number of birds eating insects and the number of prey that eat the birds (Darwin, 1985, p. 124).

Darwin's imaginative 'reflective' thinking could be observed when he was looking intently at a flower 'for ten minutes', although his gardener Lettington misinterpreted this as a sadness of mind ('[he] has been very sadly [...], he moons about in the garden [...]'). But Darwin's keen, continued observation was no doubt an examination of the flower's interrelationship of parts and their developmental history (Levine, 2011, pp. 141–2; Browne, 2003, p. 460). This 'rastro' observation can be seen as akin to Goethe's discovery of the plant archetype while gazing at plants in the botanic garden in Italy, and Darwin's seeing the history of the ocelli in the plumage of the Argus pheasant. For Darwin, the 'rastro' method of observation, like Goethe's 'Genetic Method' and the 'Humboldtian

Method', is both a key into the soul of Nature and into the soul of Mankind, uniting their common history, sharing their tapestry of colours, brush strokes, subtle shades, textures and prose.

Analysing the 'rastro' has already highlighted the importance of Darwin's poetic imagination in relating facts to help him interpret the secrets of Nature. The discussion will now examine the influence of poetry on Darwin's sensitivities to his experience of Nature, how this developed his own poetic interpretation of Nature, and finally, how he moved from an anthropocentric to an anthropomorphic view of Nature that was transformed into a selfless poetic science, one of the highest forms of objective Mind and imagination.

Darwin's children rejected Darwin's view of himself 'as a deadened, anaesthetic man' (Browne, 2003, p. 429; Levine, 2008, p. 139) and this is reflected by his own account of himself as a young man in his autobiography where he states he had a passion for poetry and literature:

> I took much delight in Wordsworth and Coleridge's poetry [whilst at university]; [...] Formerly Milton's *Paradise Lost* had been my chief favourite, and in my excursions during the voyage of the *Beagle*, when I could take only a single volume, I always chose Milton. (Darwin, 1995, p. 31)

This passion is reflected in Darwin's *Coquimbo notebook* in an entry for May 1835 in which 'Milton' forms part of a to do list: 'Spunge [...] – Blacking – Milton – Clothes Washed – Shoes blacking [...] – write letters' (Chancellor, 2009, p. 480, [132]).

Darwin shows deep regret at having lost the pleasure of literature in later life:

> This curious and lamentable loss of the higher aesthetic tastes is all the odder, as books on history, biographies, and travels (independently of any scientific facts which they may contain), and essays on all sorts of subjects interest me as much as ever they did. (Darwin, 1995, p. 51)

But Darwin is not unfeeling, as he has strong memories of poetry's pleasures (Levine, 2008, p. 135). Levine, citing Beer, argues that reading Milton sent Darwin into 'tropical raptures', feeding his imagination and encouraging his sense of 'multiplicity, profusion and abundance' (Levine, 2008, p. 141). The light of the seas off South America reminded Darwin of Milton's *Paradise Lost*, in which the wilderness equated to 'chaos and anarchy' and is noted in Darwin's *Beagle Diary* entry for 24 October 1832:

> The night was pitch dark, with a fresh breeze. – The sea from its extreme luminousness presented a wonderful & most beautiful appearance; every part of the water, which by day is seen as foam, glowed with a pale light. The vessel drove before her bows two billows of liquid phosphorus, & in her wake was a milky train. – As far as the eye reached, the crest of every wave was bright; & from the reflected light, the sky just above the horizon was not so utterly dark as the rest of the Heavens. – It was impossible to behold this plain of matter, as it were melted & consuming by heat, without being reminded of Miltons [sic] description of the regions of Chaos & Anarchy. (Keynes, 2009, pp. 142–3)

The strange 'luminousness', the water glowing 'a pale light', the 'billows of liquid phosphorus', the 'milky train', the 'bright' crests as well as the light reflected in the 'pitch dark' sky all contributed to this feeling of an unearthly 'Chaos & Anarchy' in which everything is 'melted & consuming by heat'. Here Milton's poetry has sparked Darwin's imagination[7] and fired his sensitivity to the experiences of Nature.

Darwin intermingles his thoughts and memories of Wordsworthian and Miltonic poetry with his own experiences of Nature (Levine, 2008, p. 142). The memories portrayed in Darwin's works are both the memories of his own experiences as well as a reflection of the historical narrative of the life of Nature and its beginnings through fossils, or the 'memories' of transmutations such as those expressed in the ocelli of the Argus pheasant. Like Humboldt before him, Darwin pays attention to the details of Nature and responds with his feelings to what he sees and experiences, blending the aesthetic with the scientific (Levine, 2008, p. 143). This feeling of 'Chaos & Anarchy' is frequently experienced as 'sublime' during his five-year voyage, particularly when the insignificance of Man is brought out by the backdrop of a wild mountainous landscape further intensified by the

---

7   Like Darwin, Milton also wrote from his imagination, especially following the onset of his blindness.

wild weather. This also expresses his move away from an anthropocentric view of Nature. Darwin captures this in his *B. Blanca notebook* entry for 8 June 1834 in Tierra del Fuego, where the 'authority' of Man evaporates and the sovereignty of Nature reigns supreme:

> Curious scenery constant dirty cloud driving clouds peeps of rugged snowy crags: blue glaciers: rainbows squalls – outline against the lurid sky: [...] no claims no authority here man. – How insignificant does wigwam look – [The] Fuegian man does not look like [...] the lord of all he surveys – [...] The inaccessible mountains wider power of nature despise for control seem to say here we [...] are the sovereign. (Chancellor, 2009, p. 161, [80a]–[81a]; also p. 142)

In the *Despoblado notebook* entry for 6 June 1836, Darwin again captures the insignificance of Man and the Miltonic atmosphere of 'Chaos & Anarchy' in which he defines 'perfect chaos' as the solitude and desolation of Nature, its sovereignty again reigning supreme:

> Perfect chaos = country very desolate solitary mountainous, few animals, farm houses in valleys – no trees, wild deer large white vultures like Condors – Band of mountains. (Chancellor, 2009, p. 542 [63b])

This power of Nature's sovereignty reigning supreme over Man is expressed in the *Voyage* when describing the natural beauty at Bahia in San Salvador on 1 August 1836. Nature is so powerful that 'even in the vicinity of large cities' the power of Nature is stronger than 'the artificial labour of man' through his buildings and cultivated land:

> The whole surface is covered by various kinds of stately trees, interspersed with patches of cultivated ground, out of which houses, convents, and chapels arise. It must be remembered that within the tropics, the wild luxuriance of nature is not lost even in the vicinity of large cities; for the natural vegetation of the hedges and hill-sides, overpowers in picturesque effect the artificial labour of man. Hence, there are only a few spots where the bright red soil affords a strong contrast with the universal clothing of green. From the edges of the plain there are distant glimpses either of the ocean, or of the great bay bordered by low wooded shores, and on the surface of which numerous boats and canoes show their white sails. (Darwin, 1989 p. 366)

Once again the power and sovereignty of Nature is expressed through 'the wild luxuriance of nature'. Even the softer vegetation in the form of hedges and hill-sides 'overpowers' Man's labour, and the ocean and 'the great bay' dwarf Man's existence through the insignificance in size of the boats and canoes. This insignificance of Man and his labours is further brought out by the power of the midday sun, which makes the whitewashed houses appear like insubstantial shadows rather than buildings:

> The houses, and especially the sacred edifices, are built in a peculiar and rather fantastic style of architecture. They are all whitewashed; so that when illuminated by the brilliant sun of midday, and as seen against the pale blue sky of the horizon, they stand out more like shadows than substantial buildings. (Darwin, 1989 pp. 366–7)

Man is therefore not the centre of the universe, Nature is. And Nature has not been made for Man's enjoyment, but has been created by itself for itself – that is, it has been created for everything that is part of Nature, not just for Man. It has been created for Nature's 'menagerie':

> The land is one great wild, untidy, luxuriant hothouse, which nature made for her menagerie, but man has taken possession of it, and has studded it with gay houses and formal gardens. (Darwin, 1989, pp. 367–8)

The move from an anthropocentric view of the world is reflected in Darwin's anthropomorphic descriptions of Nature in which Man is now seen as part of Nature, not the centre of Nature; Man, though more developed in terms of consciousness, shares a common consciousness and a common history with all Nature's organisms, from worms and barnacles to apes and human beings. All share a common narrative. In Darwin's *Rio notebook* dated 17 April 1832, he compares 'twiners entwining twiners' to 'tresses like hair' (Chancellor, 2009, p. 45, [27b]). This underlines the common features that organisms[8] in Nature share, enabling the same descriptions to be used.

---

8   The term 'organisms' is here used to cover both the animal and plant world as Darwin strongly believed in a continuum of consciousness from the lowest to the highest organisms, including plants. This is clearly expressed in Darwin's *Movement in Plants* (1880) describing climbing plants when he acknowledged 'consciousness' as the ability of the organism to react to its environment through its 'sense-organs': 'It is hardly an exaggeration to say that the tip of the radicle thus endowed [...] acts like the brain of one of the lower animals' (Darwin, 1880, p. 573). With this interpretation of

Darwin is able to do this through his poetic imagination by putting his mind into the consciousness of the organism. The very act of using this common language is a statement saying that all organisms share a common consciousness, although graded. This recognition must also have been reinforced by Darwin's experience of seeing the natives of South America taking on the personae of animals during various ritualistic dances, for example natives imitating the Emu bird in the Emu dance and the movements of a kangaroo in another dance (Darwin, 1989, pp. 331–2). In the *Descent* (Darwin, 2004, p. 118), Darwin attributes human traits to his dog when he describes it as a 'sensible animal' imagining a stranger when the breeze moves a parasol (like a human being inventing a God to explain a mysterious event [Levine, 2011, pp. 165–6]).

An important example of Darwin's poetic sensibilities, sitting in an intermediate stage between the influences of Wordsworth and Milton, and his own later scientific poetic prose, can be seen in his description of the ship's spider in his *Journal of Researches*.[9] It has both 'scientific rigor' and 'a sensibility that leans towards analogy and metaphor' (Levine, 2008, p. 145):

> The little aeronaut as soon as it arrived on board, was very active, running about; sometimes letting itself fall, and then reascending the same thread; sometimes employing itself in making a small and very irregular mesh in the corners between the ropes. It could run with facility on the surface of water. When disturbed it lifted up its front legs, in the attitude of attention. On its first arrival it appeared very thirsty, and with

---

'consciousness', intention can also be included on the same continuum. As stated by Jones, 'Any creature, animal or vegetable, needs, as it copes with the outside world, to find out what is going on, to pass the information to the appropriate place and to respond to the challenges presented by Nature' (Jones, 2009, pp. 166–7). In this respect, such behaviours could be seen to be expressions of conscious intention, but taking the plant's power too literally (as if it had a brain) would result in the assumption that plants were moral beings – and this question is outside the scope of this book.

9   The spider is first noted in Darwin's *St Fe notebook* on 7 October 1833: 'Saw a largish (running spider) Shoot [sic] several times long lines from tail, there by slight air not perceptible & rising current were carried up-wards & out wards (glittering in the sun) till at last spider loosed its hold, sailed out of sight the long [...] lines curling in the air' (Chancellor, 2009, p. 189, [30a]).

> exserted [sic] maxillae drank eagerly of the fluid; [...]: may it not be in consequence of the little insect having passed through a dry and rarefied atmosphere? Its stock of web seemed inexhaustible. While watching some that were suspended by a single thread, I several times observed that the slightest breath of air bore them away out of sight, in a horizontal line. On another occasion [...] I repeatedly observed the same kind of small spider, either when placed, or having crawled, on some little eminence, elevate its abdomen, send forth a thread, and then sail away in a lateral course, but with a rapidity which was quite unaccountable. (Darwin, 1989, p. 148)

The introduction of the spider as 'the little aeronaut' immediately tells the reader that this is no ordinary spider. It is more like an explorer travelling through space, akin to a modern-day astronaut, except travelling through air. It is also not dissimilar to a colonist taking over a new territory with its spinning of a 'very irregular mesh in the corners between the ropes'. With the use of 'employing itself', this very active running around feels more like a person trying to kill time by finding something to occupy him/herself than a mere spider. The lifting of its legs 'in the attitude of attention' gives it a kind of human consciousness, as if it is thinking about something. The spider does not just need drink but appears to be conscious of its lack of fluid so that it feels 'thirsty', another human attribute. From a close observation of the data before his eyes, Darwin is not only able to portray the scene poetically, but is able to make assumptions about where it has been and where it might go. Its 'thirst' suggests it has 'passed through a dry and rarefied atmosphere', which further suggests it can move from place to place through the air, able to 'send forth a thread, and then sail away', particularly as 'its stock of web seemed inexhaustible'.

Darwin describes another spider in equally poetic prose that reinforces his hypothesis that spiders can travel great distances, but also through the length of his description over two pages and the time spent focussing on the minutiae of detail, demonstrates the importance he places on such a creature's ability to shine a light on Nature's secrets:

> A spider [...] while standing on the summit of a post, darted forth four or five threads from its spinners. These glittering in the sunshine, might be compared to rays of light; they were not, however, straight, but in undulations like a film of silk blown by the wind. They were more than a yard in length, and diverged in an ascending direction from the orifices. The spider then suddenly let go its hold, and was quickly borne out of sight. The day was hot and apparently quite calm; yet under

> such circumstances the atmosphere can never be so tranquil, as not to affect a vane so delicate as the thread of a spider's web. If during a warm day we look either at the shadow of any object cast on a bank, or over a level plain at a distant landmark, the effect of an ascending current of heated air will almost always be evident. And this probably would be sufficient to carry with it so light an object as the little spider on its thread. (Darwin, 1989, p. 149)

This description of the spider creates wonder and amazement. It stands 'on the summit of a post' as if a mountaineer conquering a mountain top surveying the territory below. It produces 'threads from its spinners', 'glittering' like gems but so light as to be mistaken for 'rays of light' such that they bend with the movement of the air in 'undulations', almost insubstantial 'like a film of silk'. The description fills the reader with amazement – the threads from the spider are 'more than a yard in length'. And then there is the mystery of where the spider goes when it is 'borne out of sight', after it 'suddenly let go its hold'. But after a careful reading of the 'rastro', everything makes sense. The delicate construction of the spider and its threads gives it incredible power. For humans the day is hot, still and calm. But a careful observation will detect 'the effect of an ascending current of heated air' that would probably 'be sufficient to carry with it so light an object as the little spider on its thread'. The reading of the 'rastro', combined with the poetic descriptions of wonder, enable Darwin to think imaginatively, to create possible hypotheses from sense impressions that initially seem impossible. 'The little aeronaut' anticipates the notion that nothing is static, that organisms can change and move, leading on to Darwin's theory of natural selection, and that the whole action does not have to be seen[10] in order to make a hypothesis just in the same way as the borne-away spider does not mean it no longer exists just because one can no longer see it – the observer can imagine where it has gone and how it got there based on the springy nature of the threads and on the warm air currents.

Darwin's move from anthropocentric poetry to anthropomorphic prose can be understood when comparing how Wordsworth and Darwin dealt with grief. Both Darwin and Wordsworth needed a law maker to create a structure for their world in which to immerse themselves, away

---

10 As with Milton's 'blind' imagination already referred to.

from the feelings of grief. For Darwin, who cannot find any consolation in Nature to make the death of his daughter Annie meaningful, the law maker is Nature itself. Darwin overcomes his grief by immersing himself in his natural history work and publications. For Wordsworth's *Ode to Duty* (written in memory of his brother John who died at sea in 1805) it is the reaffirmation of his faith in a Deity that is linked to the creation of the laws of Nature; and this reaffirmation takes the form of 'duty' in recognising that Nature is infused with divinity:

> Stern Lawgiver! Yet thou dost wear
> The Godhead's most benignant grace. (Wordsworth, 2015, ll. 41–2)

Moving on from the childlike belief that love and joy can fix all ills, 'duty' is seen as the lawgiver's structure to guide him to 'humbler functions' away from the impulses of selfishness:

> Oh, let my weakness have an end!
> Give unto me, made lowly wise,
> The spirit of self-sacrifice. (Wordsworth, 2015, ll. 52–54)

In moving away from an anthropocentric self-centred view of the world, Wordsworth and Darwin can be said to have achieved a more objective art form. Darwin can also be said to be following 'duty'[11] in the Wordsworthian sense, in that by moving away from an anthropocentric world to an anthropomorphic world, his prose is moving towards a more selfless form of poetic science, that is less personal and more universal, benefitting mankind and science. Darwin's 'moral sense' of *ought* was influenced by Mackintosh, as already discussed. This is the same as Kant's 'duty',[12] quoted by Darwin in

---

[11] Darwin could be said to have been influenced by a sense of duty from both sides of the Darwin and Wedgwood families: his father Robert Darwin's sense of duty towards his patients, his uncle Jos Wedgwood sense of duty as an MP fighting for the abolition of slavery, and his wife Emma's (Jos' daughter's) sense of duty in helping the poor (Healey, 2002, p. 129 and pp. 134–6).

[12] Although in his *Critique of Practical Reason* (1788) Kant states that the two things that fill his mind are *'the starry heavens above and the moral law within'* (Kant, 1889, p. 260), it is the 'invisible self' that 'exhibits me in a world which has true infinity'; it

the *Descent*,[13] a universal 'naked law in the soul' unaffected by individual 'appetites'. Darwin has a fond memory of 'pure' poetry, but once this has been transformed into an imaginative selfless form of poetic science he is unable to go back to a state of pure innocence. Although this new transformed state produced his 'ideas', they affected his health through seeing the beauty and destruction of Nature, of having some of his children and friends taken away through death without purpose, without a God to fall back on, creating the emptiness of the sublime, knowing that poetry helped create his ideas but at the same time creating his emptiness bordering on despair – that Nature created Man and that Man was morally responsible for himself. Thus Darwin's feeling of nausea when attempting to read Shakespeare later in life.[14] Darwin's *Beagle* voyage was his Eden and his innocence was destroyed by his curiosity. But this move from his inner imaginative poetry through the experience of the wonder of science eradicated the self through concentrating on the detail, enabling him to merge with objective Nature (Levine, 2008, pp. 215–18; Stott, 2003, p. 142). Darwin needed to move from his own imaginative poetry to the objective poetry of science as imagination got in the way of science. For example, in the *Origin* Darwin initially doubted that the mechanism of the eye could be formed by natural selection:

---

is through the '*intelligence*' of the person that the moral law reveals 'a life independent on animality and even on the whole sensible world'. It is this 'reaching into the infinitude' that creates this feeling of 'awe' or wonder for Kant (Kant, 1889, p. 260).

[13] 'Immanuel Kant exclaims, "Duty! Wondrous thought, that workest neither by fond insinuation, flattery, nor by any threat, but merely by holding up thy naked law in the soul, and so extorting for thyself always reverence, if not always obedience; before whom all appetites are dumb, however secretly they rebel; whence thy original?"' (Darwin citing Kant, 1836, p. 136, in Darwin, 2004, p. 120, footnote 3). This higher value of duty is a reflection of his own class, as already discussed, as Darwin believed in a hierarchy in society and a 'hierarchy of nations' (Browne, 2003, pp. 342–3).

[14] 'As a schoolboy I took intense delight in Shakespeare [...]. I have tried lately to read Shakespeare, and found it so intolerably dull that it nauseated me' (Darwin, 1995, p. 50).

> *His reason ought to conquer his imagination*;[15] though I have felt the difficulty far too keenly to be surprised at any degree of hesitation in extending the principle of natural selection to such startling lengths (emphasis mine). (Darwin, 1985, p. 219)

Or put another way, 'Reason [...] is far more imaginative [...] than imagination' (Levine, 2008, p. 153). Reason reveals more of Nature's secrets than pure poetry. Although not through poetry, like Wordsworth, Darwin experiences the wonder of Nature through examining the minutiae of its general laws (Levine, 2008, pp. 214–15; p. 217). Science gave consolation to Darwin rather than poetry, although this science could be regarded as a kind of poetry:

> The wonder of particular details left him happiest when he was counting seeds in a pot, playing the piano for worms, germinating what he could find in bird droppings [...]. Out of such potentially messy, even sordid, matter emerges the sublime, emerges new life. (Levine, 2008, p. 168)

Most importantly for the development of science is the way in which Darwin objectifies natural history through his strategy of self-effacement that leads to the removal of Man as the centre of Nature. The great comes from the ordinary. For example, the tallest cliffs are the product of slow incremental rises; life on islands starts with grubby little insects and not palms; human moral sensibility comes from the sexual activity of peacocks and peahens (Levine, 2008, p. 220).

Finally, as already discussed in Darwin's use of the 'double movement of prose', moving through wonder, explanation and then wonder again, Darwin's 'scientific self-effacement' affirms his scientific theories by using the 'not knowing as a means to affirmation' (Levine, 2011, p. 95).[16] For

---

15 Another example of Darwin's 'double movement of prose' – that is, reason offers a rational explanation to something that initially appears both wondrous and impossible. But such a rational explanation also creates a feeling of wonder that such a thing is possible.

16 Whereas Darwin's 'double movement of prose' creates a unity of consciousness through a bringing together of wonder and explanation in the same temporal moment, Wordsworth's 'double consciousness' (exemplified in *The Prelude*) reflects a fissure between two consciousnesses belonging to two time periods, childhood and

example, in the *Origin* Darwin marvels at the idea of extinction as something to be initially dismissed, but then marvels at our presumption to understand the 'complex contingencies' of the relationship of species – in other words, how can we assume to know so much about the complexity of species to be able to assert that extinction of species is not possible? (Levine, 2011, pp. 95–6):

> The manner in which single species and whole groups of species become extinct, accords well with the theory of natural selection. We need not marvel at extinction; if we must marvel, let it be at our presumption in imagining for a moment that we understand the many complex contingencies, on which the existence of each species depends. If we forget for an instant, that each species tends to increase inordinately, and that *some check is always in action, yet seldom perceived by us*, the whole economy of nature will be utterly obscured. Whenever we can precisely say why this species is more abundant in individuals than that; why this species and not another can be naturalised [sic] in a given country; then, and not till then, we may justly feel surprise why we cannot account for the extinction of this particular species or group of species. (emphasis mine) (Darwin, 1985, p. 325)

Like the spider being borne away mysteriously by the wind, there are hidden causes; but careful observation and reading of the 'rastro' together with imagination can help form hypotheses. Just because the causes cannot be seen[17] does not mean they do not exist. The same applies to extinctions which can never be directly experienced but can, through the 'rastro' be imagined. As Darwin is at great pains to point out, '*some check is always in action, yet seldom perceived by us*'. Imagination tempered with reason is required to fill the gap in knowledge. This enabled Darwin to develop his theory of natural selection.

As already discussed, the imagination is another example of Romantic Mind coming from matter. But the common consciousness shared between

---

> manhood. The older Wordsworth cannot quite recover the former mind of his childhood. All he can experience is his present self and 'some other Being': 'The vacancy between me and those days/Which yet have such self-presence in my mind/That, sometimes, when I think of it, I seem/Two consciousnesses, conscious of myself/ And of some other Being' (Wordsworth, 1970, Book ii, lines 29–33).

17 As with Milton's 'blind' imagination referred to earlier.

Man and animals also reflects a sympathy towards organisms at all stages of development that share a common structure, common laws and a common history. This is another example of a move away from an anthropocentric view of Nature in which Man and other forms of life are viewed in a similar light. And this feeling of sympathy towards other creatures is a feeling at a higher level of development that helps cement Man's own morality. In his *Journal*, Darwin's attention to detail makes him see that life did not originate from the heavenly exotic forms of plants such as palms but from the sordid and unpleasant, such as dirt-feeding parasites, spiders and dung. When approaching the Cape de Verd Islands on 16 February 1832, the *Beagle* 'hove to' the Rocks of St Paul. Darwin noticed that the white colour

> is partly owing to the dung of a vast multitude of seafowl [and] under the blowpipe it decrepitates, slightly blackens, and emits a fetid odour [...] and its origin without doubt is due to the action of the rain or spray on the bird's dung. (Darwin, 1989, pp. 47–8)

This was so totally unexpected. The assumption was always that beautiful organisms originated before the less exotic. The idea that beauty came from filth seemed almost blasphemous (as suggested by Levine, 'in the beginning was the parasitic bug' as a replacement of 'in the beginning was the Word' would have made many Victorians feel disgust rather than awe [Levine, 2008, p. 152]). Darwin fears that

> it destroys the poetry of the story to find, that these little vile insects [such as the woodlouse living on the dung] should thus take possession before the cocoa-nut tree and other noble plants have appeared. (Darwin, 1989, p. 49)

Yet this reversal of the anthropocentric view of Nature through 'anti-poetry' creates a form of excitement and wonder for Darwin that such sordid life is responsible for the exotic forests that he finds so beautiful.[18] Paradoxically,

---

18　The same could be said about the writing of the *Origin* at Down House – the birth of this wondrous work came from a house filled with 'the nauseating smells from his experiments: plants rotting in green slime, skinned birds and animals' as well as 'the boiling of rancid ducks' and the vomit from 'retching that filled the house day and night' (Healey, 2002, p. 232). From filth came wondrous prose and science.

The Transmutation of Darwin's Romanticism

the creation of beauty from something so vile can also appear to be miraculous, but for Darwin the miracle comes from the laws of Nature and not from something Divine.

This theme of commonality between lowly beings and developed beings is emphasized in the *Origin*. Here Darwin states that their 'chemical composition', 'cellular structure' and 'laws of growth and reproduction' are the same – for a Victorian it must have been shocking[19] to imagine a human being sharing a common composition with 'vile insects', let alone a 'hairy quadruped':

> All living things have much in common, in their chemical composition, their germinal vesicles, their cellular structure, and their laws of growth and reproduction. (Darwin, 1985, p. 455)

Darwin's imaginative ability through his intense Wordsworthian observations to see the larger systems through smaller organisms, like the ship's spiders, barnacles or worms (Levine, 2008, p. 147) also helps him develop his imaginative sympathy of being able to enter the minds of animals. For example, Darwin enters the minds of female birds when working out his theory of sexual selection. His own observations as a Victorian bird watcher are mirrored as he enters the 'mind' of the hypothetical person from another planet, viewing the 'pretty girl' at a rustic fair (Levine, 2008, p. 195). Levine regards Darwin's anthropomorphism as 'zoomorphism' as 'humans are animals, and therefore one can – as an animal oneself – understand non-human behaviour simply by imagining one's way into the animal's mind' (Levine, 2008, p. 197). In experiencing the wonder of the spider, the worm or the barnacle, Darwin wants to demonstrate that the wonder comes from the organism's own activity and that the wonder does not come from the thought that this activity is due to a divine creator (Levine, 2008, p. 145).

---

19 Reference has already been made to the Victorian feeling of anathema towards Darwin's view that Man and the Apes are related. This could equally apply to worms as they are lower down the animal chain than apes. But also 'in late-Victorian society there was a chronic underestimation of its "lowest" members: a sense that what was going on underground was at once shocking and dramatically impressive', as discovered when mining (Phillips, 1999, p. 60).

That is, Darwin is not looking up to a Deity for the cause of or reverence for creation, but is looking down to the bowels of Nature itself in the form of the worm, for example. Life and death are inextricably mixed together as the earth and dead vegetable matter pass through the living worm. Darwin believed there was a continuity of consciousness between the worm and human beings: 'We can hardly escape from the conclusion that worms show some degree of intelligence'[20] (cited by Levine, 2008, p. 149). Although the worms Darwin tested in his house with piano music did not come out of their pots to show appreciation to the music, they did show sensitivity to vibration by retreating into their burrows when placed on the piano (Levine, 2008, pp. 149–50 and Irmscher, 2004, p. 103). Such reflections seem to turn Nature inside out, making it into a bundle of opposites of what one would expect. Perhaps the greatest paradox is that of 'mindlessness producing the mind that can detect it' (Levine, 2011, p. 219).

The influence of Victorian values has already been examined, but it is important to examine the extent to which the culture of Darwin's time helped shape his imagination while on the *Beagle* and upon his return to England. Of particular importance is the Victorian view of the woman's role in pleasing man and how this may have helped influence the development of Darwin's theory of sexual selection. It is perhaps paradoxical that such a view, regarded negatively by today's standards, might have contributed in the development of this theory. This cultural influence could also be seen to be wormlike in terms of the casts or 'rastros' it leaves in Darwin's imagination. According to Darwin's *Autobiography*, his choice of contemporary fiction[21] was determined by his desire for the appearance of 'pretty women':

> A novel, according to my taste, does not come into the first class unless it contains some person whom one can thoroughly love, and if a pretty woman all the better. (Darwin, 1995, p. 51)

This is similar to the view presented in Coventry Patmore's poem *The Angel in the House*, already discussed, which depicts the woman's role as pleasing man. The notion of a 'pretty woman' pleasing Darwin is another example

---

20   From Darwin's *The Formation of Vegetable Mould through the Action of Worms* (1881).
21   It might also be expected that Darwin's choice and appreciation of novels with complex plot structures would mirror his own plots and narrative structures.

of this Victorian culture. This desire for a 'pretty woman' in literature can be seen as a reflection of Darwin's own sexual desire and as a 'fall' from the elevated height of being human to the 'primitive' level of being an animal. Ironically in birds, choice is given to the female to choose the male but not in the human as women are seen as less intelligent than men and therefore men choose their women (the reversal in role is reflected by the bird feathers women wear in their hats as ornamentation). This is doubly ironic as the choice that female birds have in selecting their mates drives sexual selection, thereby creating two varieties of plumage, one for the male and one for the female (that is, dimorphism). But Darwin does not recognize this power of the female in creating dimorphism in humans; he only recognizes it in animals. However, he did once imagine such a choice by way of analogy in the *Descent* when he says that a 'pretty girl' could exert such a choice over rustics at a fair (Levine, 2008, p. 195):

> With respect to female birds feeling a preference for particular males, we must bear in mind that we can judge of choice being exerted, only by analogy. If an inhabitant of another planet were to behold a number of young rustics at a fair courting a pretty girl, and quarrelling about her like birds at one of their places of assemblage, he would, by the eagerness of the wooers to please her and to display their finery, infer that she had the power of choice. (Darwin, 2004, p. 473)

The aesthetic concept of 'pretty' is presented by Darwin as something that can only be appreciated by higher cultures, thus the 'savage's' sense of beauty is regarded as inferior to that of birds (Levine, 2008, p. 192):

> Judging from the hideous ornaments, and the equally hideous music admired by most savages, it might be argued that their aesthetic faculty was not so highly developed as in certain animals, for instance, as in birds. Obviously no animal would be capable of admiring such scenes as the heavens at night, a beautiful landscape, or refined music; but such high tastes are acquired through culture, and depend on complex associations; they are not enjoyed by barbarians or by uneducated persons. (Darwin, 2004, p. 116)

Darwin is here comparing the 'savage's' culture to that of Victorian 'high' culture, which he regards as far superior and which the Victorians felt was superior to non-European cultures. This is brought out by savages' ornaments and music being described as 'hideous' and the idea that 'high tastes' cannot be 'enjoyed by barbarians or by uneducated persons'. Thus the view supported by Darwin that missionaries should change savages' culture and make them more civilized. This view was exemplified by FitzRoy's previous voyage on the *Beagle* when he captured four Fuegians in Patagonia for stealing one of his boats and took them back to England to civilize them, giving them English names. They were Fuegia Basket (Yokcushla) aged 9, Jemmy Button (Orundellico) aged 14, Boat Memory (original name unknown) aged 20 and York Minster (El'leparu) aged 26. They arrived in England in 1830. Boat Memory died of smallpox upon arrival in Plymouth. The others went to a school in Walthamstow to be educated and to learn to sing hymns. They were shown off on the London social circuit and were presented to King William IV and Queen Adelaide. In 1831 they were taken back to Patagonia by FitzRoy on the *Beagle* when it sailed with Darwin. Disappointingly for FitzRoy, Jemmy Button went back to his old life, giving up his civilized ways (Bayliss, 2012, pp. 1–3). By today's standards, society would see such actions as a form of kidnapping, child abuse and a kind of ethnic cleansing in which the culture, language and personal identities of individuals are removed – even their names were taken away from them. But at the time of the Victorians it was quite natural to regard non-European races, as with women, as inferior. However, Levine argues that Darwin's cultural attitudes towards women were drivers in his theory of sexual selection and are not 'reflections of complicity in Victorian sexism'. He believes Darwin's cultural attitudes were driven by the concept of aesthetic taste (shared by animals and humans) (Levine, 2008, p. 178 and p. 199). If humans find the adornments of male peacocks and pheasants 'pretty', then birds must also find them 'pretty' (that humans find them 'pretty' is evidenced by women wearing feathers in their hats like Victorian birds) (Levine, 2008, p. 196). This appreciation of the 'pretty' shading of birds' feathers is also appreciated by artists (Darwin, 2004, p. 487; Levine, 2008, p. 196): 'The shading [of the ocelli on the wing-feathers of the Argus pheasant] has excited the admiration of many experienced artists' (Darwin, 2004, p. 487).

Darwin shows that this aesthetic sense derives from animal sexuality, the root of which is female choice (Levine, 2008, p. 190). The *Origin* was referred to earlier in which Darwin provided evidence that the colour and patterns of male and female butterflies originated from the female form.

Levine (2008, p. 179) refers to Beer's (1983) positive view of Darwin in which his Victorian attitude towards women is nevertheless rooted in the origins of female choice in animals, thus showing that Darwin's views are coloured by his culture but, at the same time, showing that he is able to break free from it to shape it. This is similar to the idea of Nature being both producer and creator – Darwin's attitudes have to an extent been shaped by the Victorian period in which he lived, but at the same time he was able to contribute towards its change through his theories.

These subtle origins of intention and dimorphism of the female form could be seen as harking back to the sixteenth-century ancient Egyptian goddess of Nature, Isis, at Saïs, also known as Artemis and Diana of Ephesus (see Figure 10). The many breasts in the image make people believe 'that she nourishes all animals and all living beings' (Hadot, 2006, p. 235). The primordial origin of life in Nature, like the mythological goddess of Nature, could be seen to have come from a female form, or in a biblical inversion, from Eve, with Adam coming from her rib. Although the *Descent of Man's* theory of evolutionary change is 'antagonistic to the idea that everything has a purpose', sexual selection can be seen to emanate from dimorphism which explains racial and sexual difference (Levine, 2008, p. 187). 'The female chooses among differences' and in so doing creates divergences which thus creates dimorphism in birds, and ultimately dimorphism in humans (Levine, 2008, p. 195). Nipples in men, for example, can be seen as a vestige of the female form, again reflecting the power of Isis. However, this shift in the power of choice from females in animals to males in humans was seen by Darwin as an evolutionary shift to a higher cultural level, thus supporting the cultural view at the time that man was superior to woman. This provides the link between Darwin's misogyny, Victorian racism and sexual selection. This shift in power of choice from females in animals to males in humans appears to justify the superiority of males in human society. Such a view at the same time seems to strengthen the already racist view that other groups in society, such as the Irish, are inferior to the English gentleman. Such a misguided view (by today's standards) is ironic, as seeing women as 'pretty' objects of pleasure rather than seeing them as intelligent equals, helped Darwin work out his theory of sexual selection through *female choice*.

Figure 10: *Diana of Ephesus*. This image has been reproduced with kind permission of Whetton & Grosch. Replica sculpture of Diana of Ephesus. © Whetton & Grosch, Museum Models 2016. <http://www.whettonandgrosch.co.uk>.

Although Darwin grades civilizations according to how they measure up to Victorian culture and refers disparagingly to some,[22] when visiting Tahiti in November 1835 he is nevertheless able to appreciate the beauty of the natives' form when their expressions appear less 'savage' and they show they are 'advancing in civilization':

> The Tahitians with their naked, tattooed bodies, their heads ornamented with flowers, and seen in the dark shade of the woods, would have formed a fine picture of man, inhabiting some primeval forest. (Darwin, 1989, p. 300)

In particular, Darwin appreciates the human forms set against the background of Nature itself such as the 'heads ornamented with flowers', 'the dark shade of the woods' and the 'primeval forest' which both give a naturalistic feel to the natives but also provide an allusion to their primordial past which is also part of the narrative. This naturalistic description of the natives is further emphasized when the description of the natives and Nature blur into one, in which the forms of both are like creepers entwining one another (like Darwin's 'twiners entwining twiners' referred to earlier), creating one form, one Nature, like an 'entangled bank':

> Most of the men are tattooed; and the ornaments follow the curvature of the body so gracefully, that they have a very pleasing and elegant effect. One common figure, varying only in its detail, branches somewhat like a tuft of palm-leaves from the line of the backbone, and curls round each side. The simile may be a fanciful one, but I thought the body of a man thus ornamented, was like the trunk of a noble tree embraced by a delicate creeper. (Darwin, 1989, pp. 293–4)

The tattoo 'ornaments [that] follow the curvature of the body' seem organic, as if creepers growing up a tree. The shape of one tattoo brings out the curves of the person transforming him into a palm. The 'line of the backbone' becomes the trunk of the palm with 'a tuft of palm-leaves' branching out from the base. The tattoo accentuated the shape and curves of

---

22 For example, when visiting New Zealand in December 1835, Darwin found that extensive tattooing on the natives puzzled and misled the observer, and that their 'twinkling in the eye' created the appearance of 'cunning and ferocity' (Darwin, 1989 pp. 305–6).

the man such that it was difficult to know whether he was man or plant: 'the body of a man thus ornamented, was like the trunk of a noble tree embraced by a delicate creeper'. This is again very Wordsworthian in blurring the distinction between the subjective and objective in Nature through Darwin's anthropomorphic descriptions. And 'the trunk of a noble tree' anticipates Darwin's 'tree of life', with the 'noble' hinting at its primordial progenitors. This is exemplified in Wordsworth's 'Tintern Abbey' (Wordsworth, 2015b, unnumbered page) in which he describes how his youthful memories of Nature's 'beauteous forms' have worked on his adult mind during his absence, creating deeds of kindness and lifting the burden of life from his soul, enabling him to look into 'the life of things'. So powerful are these recalled memories of Nature past that he believes that the present memories of his revisit can help his sister Dorothy when she is sad, as well as provide her with memories of her brother's love of Nature when he is dead. It is the memories of Nature rather than the form of the Abbey that suffuse the scene, creating the impression of a naturalistic Abbey[23] created by the speaker, in which God, Nature and Humankind are all linked. This is a good example of how Nature 'embrace[s] not only inanimate nature but (in the true eighteenth-century tradition) human nature also, [uniting] [...] the mental and material worlds' (Winkler, 1975, p. 168).

Darwin's use of the 'rastro', as demonstrated in the above discussions, has revealed not just a forensic examination of Nature's history, but also a focus on the details that sparked his poetic imagination, bringing the hidden elements of past time into focus showing how Nature's most developed creations came from its most humble beginnings. As Darwin showed, the first things in life on the Rocks of St Paul's were not the most exotic but were the most disgusting – beauty can come from filth. Darwin's poetic imagination was started by reading such poets and authors as Milton, Wordsworth and Humboldt, but he created his own poetic imagination once he was on the *Beagle* and able to experience Nature poetically for

---

23   The speaker is not at the Abbey but 'five miles' up the River Wye'. The distance emphasizes the imagination.

himself through his own direct experiences. Through his naturalism he was then able to create a scientific imagination on a higher plane, in which an anthropocentric and egocentric world was replaced with a world with Nature at its centre. Getting into the mind of humble creatures such as spiders and worms (and recognising that they shared a form of consciousness with Man) enabled him to do this.

Although Darwin carried the baggage of Victorian values[24] with him onto the *Beagle* and through his major works such as the *Origin* and the *Descent*, he was still able to penetrate the secrets of Nature through his imagination, and through his 'double movement of prose' he was able to fill his writing with wonder and therefore make it Romantic. At the same time as creating an imaginative view of Nature with mind at its centre, Darwin could also be seen as imposing order on Nature. As stated by Levine:

> Through the great sweep of Victorian thought, the Romantic idea that mind alone could and did impose order on nature because it alone could create it was certainly given great force by the very emptying mind out of nature that marked the movement of Darwin's representation of the world. (Levine, 2011, p. 150)

This view mirrors Wordsworth's 'Tintern Abbey' in which the narrator feels that the ability of the mind to impose order on the world is 'abundant recompense' for what has been lost:

> That time is past,
> And all its aching joys are now no more,
> And all its dizzy raptures. Not for this
> Faint I, nor mourn, nor murmur: other gifts
> Have followed, for such loss, I would believe,
> Abundant recompense. For I have learned
> To look on nature, not as in the hour
> Of thoughtless youth. (Wordsworth, 2015b, unnumbered page, lines 84–90)

---

24 These values are already expressed in embryonic form in Townsend's *Dissertation on the Poor Laws* (1786) and are a precursor to Smiles's *Self Help* (1859).

This lost youthful experience of Nature on his first visit is compensated by a more mature imaginative view of Nature on his revisit. The same development can be seen in the lost youth of Darwin's voyage that matures into his sublime science. However, present day readers who have experienced the 'aching joys' and the 'dizzy raptures' alongside Darwin during his *Voyage of the Beagle*, may miss the uplifting nature of these experiences when reading Darwin's later narratives.

CHAPTER 8

From Erasmus Darwin's Broth of Chaos to his Goddess of Nature

It is beyond the scope of this chapter to provide an in-depth analysis of the works of Erasmus Darwin, especially as well-researched work has already been carried out by King-Hele (1999), Fara (2012) and Priestman (2013). However, there are important thematic elements of Erasmus Darwin's vitalistic conception of materiality that can be identified as similar to the underpinnings of Charles Darwin's own brand of Romantic materialism and these will be examined in this chapter.

Erasmus Darwin, like Priestley, supported a vitalistic notion of materialism in which matter was not dead but had a vital force of power, a mix of the material and the soul. On a visit to the Treak Cliff Cavern in Derbyshire in 1767, Darwin experienced matter as a broth of chaos dispersed throughout the caves. After an excursion into a cave called 'the Devil's Arse', Erasmus Darwin excitedly told his friend Josiah Wedgwood:

> I have lately travel'd two days journey into the bowels of the earth [...] and have seen the Goddess of Minerals naked. (King-Hele, 1981, p. 43)

Darwin also wrote to Mathew Boulton, saying he would conduct experiments on his mineral finds (King-Hele, 1981, p. 44). For Darwin, descending through a narrow miners' passage would have been 'like entering the bowels of some great beast, the narrow walls lined with glistening mineral streams of barytes and calcite, milky white and yellow and green, like the inside of an alien ribcage' (Uglow, 2002, p. 144). One can image Darwin chipping off bits of rock and heating them in his experiments. To many miners of a religious disposition, the gaping shaft openings seemed like the gateway to hell, but for Erasmus Darwin they were a temple of mysteries, the altar of the goddess Nature; it seemed to him to be magical transformations and secrets. It held the secrets of time itself. (Stott, 2012, p. 162)

Erasmus Darwin did not only discover rocks and minerals in the caves but also fossils, such as trilobites, bivalve shells, and corals. This made him wonder whether life had 'started long ago, in the underground pools of an earth only recently lifted from the sea' (Stott, 2012, p. 163).[1]

This mix of life coming from the bowels of the earth was a mix of organic and inorganic material. With the new vitalistic conception of materiality at the end of the eighteenth century along with the discovery of gases and electricity, matter could no longer simply be seen as something non-living and solid, without life or spirit. This concept of life coming from apparent dead or inorganic matter can be seen to have had a significant influence on Mary Shelley and the writing of *Frankenstein* (1818). When staying with Percy Bysshe Shelley and John Polidori at Lord Byron's Villa Diodati in Geneva, they were each encouraged to write a ghost story and Mary told them about one of Erasmus Darwin's experiments in which dead dried organisms were made to come back to life again (Stott, 2012, pp. 185–6):

> They talked of the experiments of Dr Darwin […] who preserved a piece of vermicelli in a glass case, till by some extraordinary means it began to move with voluntary motion. Not thus, after all, would life be given. Perhaps a corpse would be re-animated; galvanism had given token of such things: perhaps the component parts of a creature might be manufactured, brought together, and endued with vital warmth. (Shelley, 2015, p. xi)

The 'vermicelli' referred to was misquoted as Erasmus Darwin had actually written *vorticellae* in his notes, in *The Temple of Nature*, in the section on 'Theory of Spontaneous Vitality'. The *vorticellae* is in fact a microscopic

---

1   He was so impressed by these shells that he added three scallop shells to his family crest with the motto '*E conchis omnia*' – 'Everything from shells'. When he had them painted on his carriage door, it caused an outrage as the shells symbolized piety and were worn by pilgrims. Also, regarding the origin of life as having come from the sea could also be seen as denying God as a first cause. Darwin eventually had the crest removed (Uglow, 2002, pp. 152–3).

aquatic filament found in lead gutters. It appears dead when dried out, but bursts into life when put into water:

> Thus the vorticella or wheel animal, which is found in rain water that has stood some days in leaden gutters, or in hollows of lead on the tops of houses, or in the slime or sediment left by such water, though it discovers no sign of life except when in the water, yet it is capable of continuing alive for many months though kept in a dry state. (Darwin, E., 2008, p. 96)

For Mary Shelley this creation of life out of matter became a hideous phantom:

> I saw the hideous phantasm of a man stretched out, and then, on the working of some powerful engine, show signs of life, and stir with an uneasy, half-vital motion [...]. [The creator] sleeps; but he is awakened; he opens his eyes; behold the horrid thing stands at his bedside, opening his curtains, and looking on him with yellow, watery, but speculative eyes. (Shelley, 2015, pp. xi–xii)

Percy Shelley, Mary Shelley and Byron had discussed Galvani's experiments of getting dissected frogs' legs to contract from electrical currents and this inspired Mary to believe that life could be created by scientists and not just observed (Uglow, 2002, p. 429):

> Perhaps a corpse would be re-animated; galvanism had given token of such things: perhaps the component parts of a creature might be manufactured, brought together, and endued with vital warmth. (Shelley, 2015, p. xi)

Mary Shelley must have been reminded of this 're-animated' corpse when years later after Byron's death in April 1824, she paid her last respects to his corpse in London after it had been brought back from Greece. The irony here is that there was a retrogressive transformation from a handsome Romantic poet to a rotting Frankensteinian-like monster. It was as if the spirit had been sucked out of his body by the removal of all of his main sense organs: his lungs, by the Greeks as a momento for his support of their cause, his eyes and his brain:

> [...] the corpse they saw was unrecognisable as his. The teeth of which he was so proud had been discoloured by the embalming fluid and there were marks of a hacksaw – or

similar instrument – on his forehead. His features had been further distorted by the removal of his brain and eyes. (Ellis, 2011, p. 158)

There is a similarity here with Charles Darwin's reference to the filthy sea fowl dung, on the Rocks of St Paul's, enabling exotic life to grow. For both Erasmus and Charles, life can be created out of filth, but for Mary Shelley horror is also born. This mix of the hideous and the beautiful can be seen in the birth of the monstrous creation from Mary Shelley's imagination, alongside the birth and death of her own children, as well as the untimely suicide of John Polidori at the age of twenty-five, the death of her husband Percy Bysshe in a boating accident at the age of twenty-nine, and the death, due to a fever, of Lord Byron at the age of thirty six.

But for Erasmus Darwin, metamorphosis was not a phantom but was very much a part of life, whether this be the forces of 'repulsion' or 'attraction' creating the universe (Priestman, 2013, p. 23), the butterfly coming from the caterpillar, or animals changing their characteristics in order to adapt to their surroundings and avoid extinction (Uglow, 2002, p. 429):

> Shout round the globe, how reproduction strives
> With vanquished Death – and Happiness survives;
> How Life increasing peoples every clime,
> And young renascent Nature conquers Time. (Darwin, E., 2008, Canto IV, lines 451–54)

This notion of creation from death and rebirth is captured in part one of *The Botanic Garden*, in *The Economy of Vegetation* (1791), in which Erasmus Darwin draws on William Herschel's early papers on the 'Construction of the Heavens' (1785 and 1789), describing an evolving universe with 'distant nebulae growing and expanding like plants' (Holmes, 2009, p. 197). This emphasizes the possibility of the evolution of the universe over unimaginable time, putting 'the Creator at an increasing shadowy distance from his Creation' (Holmes, 2009, p. 92). This blend of birth and death is captured enigmatically in

> Roll on, YE STARS! Exult in youthful prime,
> Mark with bright curves the printless steps of Time. (Darwin, E., 2010, Canto IV, lines 367–8)

Here Time itself is printless because time cannot actually be seen, yet the effects of time in the universe can be seen as 'bright curves' indicating the birth and death of stars, or on earth the hieroglyphs of the caves made up of fossils and minerals in various stages of birth and decay. For Erasmus Darwin, as with Humboldt and Charles Darwin, the heavens and life on earth make up the wholeness and the unity of Nature, and this is reflected in

> Flowers of the sky! Ye too age must yield,
> Frail as your silken sisters of the field! (Darwin, E., 2010, Canto IV, lines 371–2)

In his notes to *The Economy of Vegetation*, Darwin explains 'Near and more near' in line 369 as referring to [Herschel's conclusion] 'that the nebulae or constellations of fixed stars are approaching each other, and must finally coalesce in one mass. Phil. Trans. Vol. LXXV' (Darwin, E., 2010, p. 189). This forming together of one mass creates the notion of the death of the individual stars:

> Headlong, extinct, to one dark centre fall,
> And Death and Night and Chaos mingle all! (Darwin, E., 2010, Canto IV, line 376)

Yet from this death rises the birth of new Nature, like a resurrection or a phoenix rising from its own ashes. This is expressed in 'Till o'er the wreck', in line 377, and explained in the notes: 'The story of the phenix [sic] rising from its own ashes with a twinkling star upon its head, seems to have been an ancient hieroglyphic emblem of the destruction and resuscitation of all things' (Darwin, E., 2010, pp. 188–9). Although the outward forms of Nature change, their underlying essence is the same, as Nature is immortal. This conforms to Goethe and Charles Darwin's use of the concept of 'archetype'. As with Lazarus and the phoenix, the forms rise from the ashes of death into a different life form. The Argus pheasant can be seen to represent these changing or, for Charles Darwin, evolving forms over time. Although death and destruction seem to take over, Nature lives through this chaos and survives.

This immortality of Nature and matter can be seen in *The Temple of Nature* (1803), published the year after Erasmus Darwin's death:

> Organic forms with chemic changes strive,
> Live but to die, and die but to revive!
> Immortal matter braves the transient storm,
> Mounts from the wreck, unchanging but in form. (Darwin, E., 2008, Canto II, lines 41–44)

Following the doctrine of St Paul, Erasmus Darwin equates the phoenix nature of 'the wrecks of death' to 'the resurrection of the body in an incorruptible and glorified state, with consciousness of its previous existence' (Darwin, E., 2008, p. 82). This image of the resurrection is similar to the painting of Lazarus and Charles Darwin's description of the Argus pheasant.

In the same way as the seemingly dead microscopic aquatic filament of *vorticellae* can burst into life, so too can the birth of rational Mankind appear to come to life from animal Nature, which in turn came from a material inorganic lifeless Nature. Just as Charles Darwin celebrated the development of Man in *The Descent of Man*, so did Erasmus Darwin celebrate the development of Man brought about through the industrial revolution. In *The Loves of the Plants* (1791) he portrays Richard Arkwright's spinning machine as speedier than the human hand:

> With wiry teeth *revolving cards* release
> The tangled knots, and smooth the ravell'd fleece. (Darwin, E., 2007, Canto II, lines 95–6)

But despite this celebration of development for the benefit of Mankind, there is a deep contradiction in the way it is celebrated, as the individual worker is forgotten in the manufacturing process and seems to remain as an undeveloped mass of physical Nature, a mere accessory in relation to machinery. In *The Temple of Nature*, for example, there is no mention of the worker in the spinning process:

> ARKWRIGHT taught from Cotton-pods to cull,
> And stretch in lines the vegetable wool;
> With teeth of steel its fibre-knots unfurl'd,
> And with the silver tissue clothed the world. (Darwin, E. 2008, Canto IV, lines 261–4)

As pointed out by Priestman:

> We are left with the idea that, through the steel teeth of his machines rather than his labour force, and then through his trading networks rather than the sailors and dressmakers they employed, Arkwright spun the cotton and 'clothed the world' himself', highlighting Raymond Williams's identification in the shift in meaning from 'industrious' labourer to the 'industrial' output of machines and managers. (Priestman, 2013, p. 96)

As a doctor, Erasmus Darwin would have been aware of the dangers of metal dust leading to consumption or the dangers of the use of lead in glazing. Yet there is no evidence that he discussed these issues with his medical friends. The benefits of industry were no doubt seen as outweighing their dangers (Priestman, 2013, p. 97). Despite this neglect to praise the achievements of the individual worker, there is very much an acknowledgement of Man's inseparable place in Nature. The prose helps the reader imagine the workers as accessories to the spinning machines, imagine them as almost cogs themselves, becoming absorbed by the machinery, almost turning into machines themselves, helps the reader to experience the very nature of Nature, entering the industrial equivalent of the 'bowels of the earth' or the broth of chaos, smelling and tasting the mix of the sweat of the workers mixed with the steam and the dust of the fabric; visualising human bodies moving in time to, and being shaped by, the motion of the machinery. This naturalizes the industrial process. The imagery mirrors Humboldt's concept of one integrated Nature and Charles Darwin's 'web of affinities'. In *The Economy of Vegetation* (1791), Erasmus Darwin emphasized this interdependence of Nature through the dependence of animals' survival on plants – an 'economy' or structure and organization of Nature akin to the circulation of blood in the human body, discovered by Harvey and developed by the Physiocrats in their understanding of the economy of working the land. Circulation and interdependence along with technological development may be seen as improving some aspects of the individual and the nation's health and well-being, whether this be political, biological, ethical, educational or financial (Fara, 2012, pp. 148–9). Erasmus Darwin illustrates this interdependence in the industrial age through the imagery of earth, air, fire and water. He refers to Watt's steam engine that will transform travel and industry:

> Bade round the youth explosive STEAM aspire
> In gathering clouds, and wing'd the wave with fire. (Darwin, E., 2010, Canto I, lines 255–62)

And to Boulton's steam-powered coin-pressing machine, originally hand pressed:

> Descending screws with ponderous fly-wheels wound
> The tawny plates, the new medallions round. (Darwin, E., 2010, Canto I, lines 283–8)

Charles Darwin was almost obsessed with worms and their central role in life and death in the recycling of dead matter, forever churning the soil round and round in his garden. So too with his grandfather, celebrating the essence of life (Being or ENS,) in the form of clay found in the earth, and transformed into pottery by Wedgwood. On a grander scale, this change and development of the earth is seen in the continents heaved up above the ocean, marine shells producing limestone rocks, and minerals being produced from decomposed matter. For Erasmus Darwin this perpetual circulation of vegetable matter and animal bodies is the essence of being or ENS, but unlike in Charles Darwin, this circulation incorporated the transmigration of spirits. There are similarities here, in an abstract sense, to Goethe's concept of archetypes that exist as unchanging forms but appear different in their outward individual states through time:

> With fostering hand the parting atoms catch,
> Join in new forms, combine with life and sense,
> And guide and guard the transmigrating Ens. (Darwin, E., 2010, Canto II, lines 572–4)

In his notes, Erasmus Darwin explains the 'transmigrating Ens' by referring to the ideas of Pythagoras:

> Hence these philosophers have supposed, that both matter and spirit are equally immortal and unperishable; and that on the dissolution of vegetable or animal

organisation, the matter returns to the general mass of matter; and the spirit to the general mass of spirit, to enter again into new combinations, according to the original idea of Pythagoras. (Darwin, E., 2010, p. 116)

'New combinations' are formed, but come from 'immortal and unperishable' matter and spirit, an unchanging essence or set of archetypes that underpins all tangible particular forms. And this transmigration of spirits from regenerating matter could be seen and felt through Priestley's discovery of pure air, used to improve the diving bell:

> Huge SEA-BALLOONS beneath the tossing tide;
> The diving castles, roof'd with spheric glass,
> Ribb'd with strong oak, and barr'd with bolts of brass,
> Buoy'd with pure air shall endless tracks pursue. (Darwin, E., 2010, Canto IV, lines 195–02)

Here too is a hint of Charles Darwin's 'rastros' yet to come, in 'endless tracks pursue', in which the diving bells not only leave their transitory surf of bubbles behind them (as 'printless steps of time') as they move through the water, but also leave a history of their movements through their records of underwater discoveries.

In 1827, at the time that Charles Darwin was a medical student at Edinburgh University, Alexander von Humboldt was visiting England in an attempt to persuade the East India Company to let him explore India, unsuccessfully as it turned out (Wulf, 2015, p. 182). While in London, Humboldt was invited by Isambard Kingdom Brunel to inspect work on the construction of the first tunnel under the river Thames. On 26 April they were lowered into the river in a two-ton diving bell, dropping to a depth of thirty-six feet. The descent must have seemed akin to Erasmus Darwin's cave experience of 'the bowels of the earth', with the murky waters showing through the two thick glass windows, and the eerie darkness punctuated by the weak glimmer of the lanterns. And just like Erasmus Darwin's 'Huge sea-balloons ... Buoy'd with pure air', so was Brunel's diving bell supplied with air from a leather hose (Wulf, 2015, p. 184). Here too Humboldt experienced the oneness of the interrelationship between Man and Nature, through the underwater pressure making his nose and throat bleed, just as climbing the heights of Chimborazo had done (Wulf, 2015, p. 185). The water swirling around the diving bell is another essence of life that can also be seen as a kind of spirit, as it can change into steam, and therefore become a vapour of gas, and then condense again as rain, as well as becoming as solid as ice. This circulatory nature of water, mirroring Harvey's circulation of blood and Quesnay's generation of wealth through the working of the land, could be seen in the use of ornamental fountains:

> NYMPHS! YOUR bright squadrons watch with chemic eyes
> The cold-elastic vapours, as they rise;
> With playful force arrest them as they pass,
> And to *pure* AIR betroth the *flaming* GAS. (Darwin, E., 2010, Canto III, lines 201–4)

The air is Priestley and Keir's 'dephlogisticated' air, otherwise known as oxygen by Lavoisier, and the gas is Lavoisier's hydrogen (Darwin, E., 2010, p. 141).

Erasmus Darwin and Charles Darwin were aware that the progress of industrialization had an impact on the welfare of its workers and fuelled the inhumanity of the slave trade. For both Erasmus and Charles they shared a similar concept of 'sympathy' for humanity and it is therefore useful to compare their views and identify any influences Erasmus might have had on Charles in this and any other respect. Industrialization made Britain rich, and the slave trade triangle linking Britain, Africa and America through the trade of slaves, gold, cotton, palm oil and sugar resulted in slaves being regarded as accessories to machinery (akin to the imagery of white 'wage-slaves' in the labour process), with the leg clamps and padlocks used on slaves as part of this industrial process, squeezing out the last vestiges of humanity, making the entrapped bodies part of the ironwork that enslaved them (Fara, 2012, pp. 168–77). Erasmus Darwin regarded this slavery as barbaric:

> E'en now in Afric's groves with hideous yell
> Fierce SLAVERY stalks, and slips the dogs of hell;
> From vale to vale the gathering cries rebound,
> And sable nations tremble at the sound! (Darwin, E., 2007, Canto III, lines 443–6)

He also believes that those who do nothing about it are as guilty as the slave traders:

And when 'inexorable CONSCIENCE hold his court' we are told that those who do nothing are also guilty: 'HE, WHO ALLOWS OPPRESSION, SHARES THE CRIME'. (Darwin, E., 2007, Canto III, lines 452–59)

However, Fara argues that Charles Darwin's Wedgwood grandfather's production of the pottery medallion with the inscribed plea 'Am I not a man and a brother', under the image of a kneeling handcuffed slave, is not a call for equality, as, at this time, brothers within a family were not equal. Fara argues that the true sentiment reflected by British abolitionists was that:

> Britons were more concerned with their own moral salvation than with the fate of others: whether it concerned Africans, animals or women, those who treated inferior beings humanely could feel confident about their personal status as civilized Europeans. (Fara, 2012, p. 178)

According to Fara, the kneeling slave can be interpreted as demanding good treatment because he is a fellow human being, 'but he is not claiming the right to equality (Fara, 2012, p. 200). Everyone knew their place in Georgian society and, like a father looking after his children, 'the privileged were responsible for looking after those born into lower stations' (Fara, 2012, p. 180). But at the same time, the privileged humanitarian freethinkers lined their pockets as a direct result of this privilege and slavery. Arkwright's looms produced clothes for slaves and Wedgwood profited from the sale of his snuffboxes with the kneeling slave motif (Fara, 2012, p. 183). Nevertheless, Erasmus Darwin did, despite these financial benefits, fight for the end of the torture of slaves and wanted their bad treatment and the instruments of their torture to be widely publicized, as expressed in this letter to Wedgwood on 13 April 1789, which Charles Darwin quotes in his biography of his grandfather:

> I have just heard that there are muzzles or gags made at Birmingham for the slaves in our islands. If this be true, and such an instrument could be exhibited by a speaker in the House of Commons, it might have a great effect. Could not one of their long whips or wire tails be also procured and exhibited? But of our own manufacture would have a greater effect, I dare say. (King-Hele, 2003, p. 87)

The inevitable contradiction arises in which Man develops civilizations through the progress of industrialization, yet neglects the welfare of its workers and permits the inhumanity of the slave trade. Yet as the work of the abolitionists shows, humanity eventually catches up with itself. These sentiments take time to develop in Man, and can be seen to be a part of the continuum of consciousness progressing from the more primitive organisms to the rational in Man's development of language, the emotions and sympathy, which is at the heart of Charles Darwin's *Descent of Man*. Erasmus Darwin develops this theme in Canto III of *The Temple of Nature*:

> To express his wishes and his wants design'd
> Language, the *means*, distinguishes Mankind. (Darwin, E., 2008, Canto III, lines 437–8)

He emphasizes the development of the emotions in 'And soft emotions mark the feeling mind' (Darwin, E., 2008, Canto III, lines 466) and in:

> The Seraph, SYMPATHY, from Heaven descends,
> And bright o'er earth his beamy forehead bends;
> On Man's cold heart celestial ardour flings,
> And showers affection from his sparkling wings. (Darwin, E., 2008, Canto III, lines 467–70)

It is quite remarkable that Charles refuses to acknowledge any influence from his grandfather, especially when both Charles and Erasmus describe animals in human terms, seeing both existing on the same consciousness continuum:

> The Wasp, fine architect, surrounds his domes
> With paper-foliage [...]
> The cunning Spider with adhesive line [...]
> Conscious of change the Silkworm-Nymphs begin [...]
> They link the reasoning reptile to mankind!
> -Stoop, selfish Pride! Survey thy kindred forms,
> The brother Emmets, and thy sister Worms! (Darwin, E., 2008, Canto III, lines 411–34)

The reference to the wasp's domes brings to mind Charles' description of the bees' hives, and 'the cunning Spider' reminds the reader of the spiders on the *Beagle*. And of course, Erasmus also mentions one of Charles' favourite creatures, the worms.

One of the most important developments in the insect world is that of touch. This is referred to in Erasmus Darwin's footnote to 'the Wasp' and this way of writing in prose may have had an influence on his grandson Charles (as much of Erasmus's poetry is in prose in his footnotes):

> Hence the excellence of the sense of touch in many insects seems to have given them wonderful ingenuity so as to equal or even excel mankind in some of their arts and discoveries; many of which may have been acquired in situations previous to their present ones, as the great globe itself, and all that it inhabit, appear to be in a perpetual state of mutation and improvement. (Darwin, E., 2008, footnote 35, p. 62)

Erasmus Darwin elaborates on the importance of touch in the development of Man's emotions and sympathies through the development of the hand. The hand not only enables humans to grasp objects but enables Man to gain a greater understanding of his environment through the delicacy of touch:

> Nerved with fine touch above the bestial throngs,
> The hand, first gift of Heaven! To man belongs;
> Untipt with claws the circling fingers close,
> With rival points the bending thumbs oppose,
> Trace the nice lines of Form with sense refined,
> And clear ideas charm the thinking mind. (Darwin, E., 2008, Canto III, lines 121–6)

This delicacy of touch, along with sight, enables the mind to conceptualize the word, creating the ability to imitate it through the re-creation of sense-impressions, language and the creative arts 'and thence to the sympathetic mirroring of others' feelings which produces socialized behaviour' (Priestman, 2013, p. 127). Sympathy is therefore the final stage of human development in which 'people bond together in societies' (Fara, 2012, p. 241). In Charles' biography of his grandfather, he emphasizes the importance Erasmus placed on sympathy (as also expressed by Wordsworth in 'Love of Nature Leading to Love of Mankind' in *The Prelude*), describing it as 'the foundation of all our social virtues':

> He urges that sympathy with the pains and pleasures of others is the foundation of all our social virtues; and that this can best be inculcated by example and the expression of our own sympathy. 'Compassion, or sympathy with the pains of others, ought also to extend to the brute creation ... to destroy even insects wantonly shows an unreflecting mind, or a depraved heart'. (King-Hele, 2003, p. 42)

As 'sympathy' for Charles Darwin was not just the 'foundation of all our social virtues' but was also the foundation of his *Descent of Man*, it is odd that he did not make any link between his grandfather's use of 'sympathy' and his own use of the word. This would have helped establish whether or not Charles' grandfather had influenced him in its use.

For Erasmus Darwin, this continuum of development leading to the social virtue of sympathy found in Man goes right back to the beginnings of life emanating from the broth of chaos, creating a vision of the creation of the universe from the explosion of matter rather than from the God of Genesis (Priestman, 2013, p. 106), presaging the 'Big Bang' theory' (Uglow, 2002, p. 426):

> When LOVE DIVINE, with brooding wings unfurl'd,
> Call'd from the rude abyss the living world.
> 'LET THERE BE LIGHT!' proclaim'd the ALMIGHTLY LORD,
> Astonish'd Chaos heard the potent word;-
> Through all his realms the kindling Ether runs,
> And the mass starts into a million suns. (Darwin, E., 2010, Canto I, lines 101–6)

But this 'LOVE DIVINE' is not the divine love of God as Divine Creator. Similar to Charles Darwin's laws of Nature coming from Nature, Erasmus Darwin's creation comes from the material depths of the 'egg of Night' floating in Chaos. This is made clear in his footnote to 'When LOVE DIVINE' in line 101:

> From having observed the gradual evolution of the young animal or plant from its egg or seed; and afterwards its successive advances to its more perfect state, or maturity; philosophers of all ages seem to have imagined, that the great world itself had likewise its infancy and its gradual progress to maturity; this seems to have given origin to the very antient [sic] and sublime allegory of eros, or Divine love, producing the world from the egg of Night, as it floated in Chaos'. (Darwin, E., 2010, pp. 16–17)

According to Uglow, this secular version of the Creation gives another meaning to 'Divine Love' as the lines 'brooding wings unfurl'd' and 'rude abyss' are 'imbued with sex' (Uglow, 2002, p. 427). The concept of Love or Sex being the same as God, and being the driving force behind the reproduction of life (Fara, 2012, pp. 224–5), refers back to the 'wings outstretch'd' and the 'bursting egg of Night' in earlier lines:

> IMMORTAL LOVE! Who ere the morn of Time,
> On wings outstretch'd, o'er Chaos hung sublime;
> Warm'd into life the bursting egg of Night,
> And gave young Nature to admiring Light! (Darwin, E., 2008, Canto I, lines 15–20)

'Sexual Love' is even more strongly expressed in the prayer to sexual enjoyment, celebrating the marriage of Cupid and Psyche. According to Fara, Darwin's 'Garden of Eden is an erotic playground, the site not of temptation but of jubilant fornication' (Fara, 2012, p. 225):

> Behold, he cries, Earth! Ocean! Air above,
> And Hail the DEITIES OF SEXUAL LOVE!
> All forms of Life shall this fond Pair delight,
> And sex to sex the willing world unite. (Darwin, E., 2008, Canto II, lines 243–46)

These natural forces of 'quasi-sexual energy' were to be celebrated for their 'role in perpetuating life and transmitting improvements to subsequent generations', personifying 'Nature as a generative sustaining goddess' (Fara, 2012, p. 62). This interest in the 'quasi-sexual energy' of Nature is reflected in Erasmus Darwin's translation of Linnaeus' botanical work *Systema Naturae* (1735) as *A System of Vegetables* (1783–1785) and *The Families of Plants* (1787), in which he reinstated the original sexual terms that William Withering had sanitized in his translation *A Botanical Arrangement of all the Vegetables Naturally growing in Great Britain* (1776) (Fara, 2012, p. 72). Erasmus Darwin's interest in sexual energy is also reflected in his use of 'camera obscura' in his *Loves of the Plants* whose Latin meaning 'darkened room' expresses Darwin's sexual metaphor of 'erotic delights [lying] ahead' (Fara, 2012, pp. 82–3). In the eighteenth century, the term *gardening* had sexual connotations. Wealthy gentlemen with large gardens used the opportunity to seduce women in their groves, temples and grottoes. The terms they used to describe their gardens, such as 'Secret Satisfaction', 'Recesses', and 'Penetration' could also be applied to their wives or mistresses (Fara, 2012, p. 68). Erasmus Darwin's own enchanted garden was developed through his relationship with a dedicated gardener, Mrs Elizabeth Pole, with whom he was besotted, and he later married, after the death of her husband. This sexual interest in botany was mirrored by Joseph Banks, and by Captain James Cook's crew during their voyage to Tahiti (1768–1771), in which they went native, following local 'promiscuous' customs. One of the voyage's main purposes was to observe the Transit of Venus, in order to measure its progress as it appeared to cross the sun. In Canto IV of *The Loves of the Plants*, Erasmus Darwin imagines 'Venus's pleasure at a "promiscuous" marriage between a hundred flowery grooms and brides'. This 'describes a nature who is fecund and polygamous, ruled by her own laws and not by those of a restrictive God' (Fara, 2012, p. 88):

> A *hundred* virgins join a *hundred* swains [...],
> Each smiling youth a myrtle garland shades,
> And wreaths of roses veil the blushing maids [...]:
> -Thick, as they pass, exulting Cupids fling
> Promiscuous arrows from the sounding string. (Darwin, E., 2007, Canto IV, lines 467–76)

Although from our present-day perspective this view of sexual energy may seem to reflect a patriarchal sexual power in which man dominates woman, Erasmus Darwin viewed the female figure as representing the Goddess of Nature, or fertility, and therefore imbued with the power of creation. This can be seen in *The Temple of Nature*'s opening frontispiece, by Fuseli, of 'a woman displaying the aristocratic profile of a Greek statue [drawing] back a veil to reveal an extraordinary female figure with three breasts' (Fara, 2012, p. 218), which relates to the Goddess of Nature or fertility (Diana of Ephesus or Isis; see Figure 10). As expressed in the poem:

> Tower upon tower her beamy forehead crests,
> And births unnumber'd milk her hundred breasts;
> Drawn round her brows a lucid veil depends. (Darwin, E., 2008, Canto I, lines 131–33)

Although there may not be any direct influences here, the power of the female form expressed through Erasmus Darwin's Goddess of Nature, and 'the deities of sexual love', may well have stimulated Charles' imagination when developing his theory of Sexual Selection. Charles Darwin was familiar with his grandfather's works and he had written a biography on him. So even if he was not directly influenced by his grandfather, it is extremely likely that his grandfather's ideas stimulated Charles' own imagination into focussing on the issues of Nature during his day. Although Erasmus Darwin's poetry was of the form and style of neo-classical poetry of the eighteenth century, his footnotes are more akin to the prose of Charles Darwin and the prose in Wordsworth's *Lyrical Ballads* which was produced as a reaction against the ornateness of poetry, which Wordsworth believed distanced poetry from the ordinary 'man'. In this respect, therefore, Charles Darwin may have been influenced by the form of his grandfather's prose as well as his poetry.

CHAPTER 9

The Rime of the Ancient Naturalist

There is no real conclusion to the Romanticism of Darwin's life, ideas or his works as they continue to live on as life forever unfolds, particularly in the imaginative world in which science and poetry blend together; a spiritual blend which is a blend of the everyday and the supernatural. The irony here is that despite Darwin's agnostic tendencies towards the First Cause of Creation, the whole narrative of the story of Creation and its origins overlapping the narrative of his own life story, has a mystical feel to it. And this is made even stronger when both narratives interact with the narrative of the reader's own life history and imagination.

This chapter, by way of conclusion, deals with the force of this mystical or supernatural view which is beautifully, if not sublimely, drawn out by Ruth Padel's *Darwin: A Life in Poems*. Ruth Padel is Charles Darwin's great-great granddaughter and so this poetical biography is a direct descendant of Darwin's own 'rastros'. Darwin's 'rastros' directly link up with Padel's own 'rastros' and these can be seen in her descriptions of her conversations with her grandmother, Nora Barlow, Charles Darwin's granddaughter, who had published the unexpurgated version of Darwin's *Autobiography*. In particular, Padel remembers her ninety-five-year-old grandmother telling her about her own conversations with Charles Darwin and how he felt his ideas had 'affected the faith of his wife, Emma' (Padel, 2010, pp. xiii–iv). There is even a link between these 'rastros' and my own 'rastos'. I purchased this book second hand and upon turning the first page discovered Ruth Padel's signature with the words 'For Victoria with best wishes and thank you for coming', dated 20th July 2013 (who was Victoria, what was the event and why did she, or her family/friends sell such a wonderful book – is there another sad story or 'rastro' behind this?). To obtain this signed copy made me feel an even stronger link to those footsteps of the past, both of Darwin's own life and the deep past of fossils and rock formations.

This mystical feeling of connection and the sublime was similar to my own personal experience when visiting the Manuscripts Reading Room at Cambridge University Library to view Darwin's *Old and Useless Notes*, forming a tapestry of 'rastros'. The notes consisted of torn scraps of paper pasted in a beige folio alongside various letters. Opening the volume was like prising open a skull, secretly peering at hidden thoughts, footprints of the mind. And amongst them all was the word 'Sublime' written in Darwin's handwriting in bold ink, with a thick, almost slimy, looking 'S' like a Coleridgean water-snake from *The Rime of the Ancient Mariner*. Physically and spiritually, this was a fossilized piece of Darwin.

So too with Ruth Padel's book. Like *The Rime of the Ancient Mariner*, it is a rhyme, but in this case it is 'The Rime of the Ancient Naturalist'. It is the story of his life but also the story of time. It creates the impression of a conversation between Padel and her great-great grandfather as well as between them and their readers. The collection of poems also needs to be understood within the context of the bicentenary celebrations of Darwin's birth held in 2009. Twenty of the poems were commissioned by the Bristol Festival of Ideas and others by the Contemporary Arts Programme at the Natural History Museum, London, for an exhibition (Padel, 2010, p. xiii). The form of the book and its poems is similar to Coleridge's *The Rime of The Ancient Mariner* (1817). Coleridge's poem is written in three voices: after every stanza or so there is a gloss or annotation written in the margin 'in the style of a seventeenth-century editor of an older text' (Scofield, 2003, note 1, p. 302);[1] the voice of the narrator in the poem; and the conversation between the Ancient Mariner and the Wedding Guest. In Padel's book the verses are annotated in the margin giving the historical context with dates. The verse is made up of a combination of the voice of Padel as the poet, expressing her own emotional interpretation of her grandfather's life, along with quotations from Darwin's 'books, scientific papers, autobiography, journals, notebooks, drafts and letters' (Padel, 2010, p. xviii). Padel's 'ras-

---

1   The poem was originally published in 1798 in Wordsworth and Coleridge's *Lyrical Ballads* in which the spelling was deliberately archaic as Coleridge wanted to explore the medieval origins of Romanticism. The marginal glosses were added in 1817. The edition referred to here is that of 1912 which follows the revised text of 1834 (Scofield, 2003, p. 302).

tros', as with Darwin's and the Ancient Mariner's, is about piecing together a history, a collection of broken memories, and trying to fill in the gaps between 'the broken trails':

> You need people who know
>   the broken trails, sudden pits underfoot,
> and the animals. Capybara, jaguar, agouti. (Padel, 2010, p. 38)

Although the *Mariner* deals with the supernatural on a religious level, there are mystical elements in both the *Mariner* and Darwin's works that share a common theme, namely the workings of the moral universe and the human mind. They also cover the working through of guilt and repentance. For the Ancient Mariner, it is about making up for the wrong of killing one of God's creatures, the Albatross, and the consequent death of his fellow sailors. For Darwin it is the feeling of guilt in developing ideas contrary to the religious beliefs of his day, God's laws of Nature replaced by the material laws of Nature, as well as the strongly held religious beliefs of his wife, Emma. In a letter to his friend Joseph Dalton Hooker, on 11 January 1844, he describes this admission of these ideas as 'like confessing a murder' (Burkhardt, 2009c, p. 2). This can be seen as not unlike the Ancient Mariner's killing of the Albatross, with the guilt of its death hanging around his neck until the spell of this guilt could be released by prayer. For Darwin this guilt could not start to be released until his 'scientific' community and the Victorian society at large could start to accept his theory of transmutation and natural selection. It took a long time for his Albatross to fall from his neck resulting in considerable personal turmoil over the years. Darwin's vicarious theological guilt could be said to have comingled with the feeling of guilt associated with the possibility of creating sickly children through marrying his cousin, Emma Wedgwood. The guilt, the vomiting, the perpetual sickness and the deaths of many members of his family and friends could well have been felt by him as a punishment for his actions, if not from God then from Nature herself. Although the reference to the metaphorical 'wedges' breaking up the face of Nature are powerfully pessimistic in Darwin's *Origin*, Darwin's works are less of a nightmare compared to the extremes of malign and beneficent forces of Nature in the *Mariner* such as the sun, the moon or the wind (the absence of wind or gale force storms could both lead to death for sailors in sail-powered vessels). But in both the works of Darwin and Coleridge, the turmoil can be seen as reflecting the external and internal states of the mind, creating 'a feeling of "one life" working through them both' (Scofield, 2003, p. xiii). Ultimately for the Ancient Mariner, recognising the beauty of the humbler elements of Nature such as the water-snakes (although for some seen as disgusting) fills him with love leading to his transformation into a better person and to his forgiveness from God:

> Oh happy living things! No tongue
> Their beauty might declare:
> A spring of love gushed from my heart,
> And blessed them unaware [...]
>
> The self-same moment I could pray;
> And from my neck so free
> The Albatross fell off, and sank
> Like lead into the sea. (Wordsworth, 2003, lines 282–91, p. 216)

This recognition of the beauty of the water-snakes is not unlike Darwin's recognition of the beauty of worms and barnacles and a move away from an anthropocentric to an anthropomorphic view of Nature.

Unlike the Ancient Mariner, Darwin's appreciation and intoxication of the beauty of Nature comes well before the development of his murderous idea. Padel is able to penetrate deep under Darwin's skin, able to express the delight of his feelings even better than he could, as if she had absorbed and articulated his own memories making them become her memories. These 1832 memories of Bahia are metamorphosed to humanize them, to make sense of them, to make them part of his mental and physical world, both past, present and future. Padel describes his reaction to the parasitic tree ferns he's never seen before as:

> Jade wagon wheels
> eighty feet up like jugglers' saucers on a pole. (Padel, 2010, p. 32)

They make you think of a circus with jugglers but with saucers impossibly high up as to be unimaginable and a size too big for juggling, more suitable for a waggon train out of a western. Everything is just ridiculously oversized and difficult within the context of being human to take in, to understand. The complex mixture of religious references, the primordial world, humanized forms and a kind of underworld of past memories, makes the experience especially mystical, even more than Darwin himself expressed:

> Bristle of orchid leaves on every black branch
>   like green flames over Bibles
> Botanical forms gyrate and pour
> through rivers of otherworld bark
>   and a wrestling musculature of pure
> live wood. This church breathing dark –
> surely he lived here as a boy, or before he was born.
>   He's a revenant, a sleepwalker. The lavender light
> is dim, like underwater. (Padel, 2010, p. 32)

There is a good mix of the everyday and the supernatural as in the intended mix of Wordsworth and Coleridge's poems in *Lyrical Ballads*. Reading this the reader could mistakenly imagine they were reading excerpts from Dylan Thomas' *Under Milk Wood* with the 'black branch', 'green flames over Bibles' and 'church breathing dark', similar to Thomas' 'small town, starless and bible-black' and the 'silent black, bandaged night' (Thomas, 1954, pp. 1–3). Like Padel's description of the 'rivers of otherworld bark' and sleepwalking as if in an underwater 'lavender light', Thomas' world is a similar one in which memories can be heard as dreams (Thomas, 1954, pp. 1–3). Darwin is himself 'walking into every dream he's ever seen' with a mix of the mystical such as 'trees like giant columns of a fallen temple', the more tangible 'slashing raindrops, plump as eggs' and the fairy-tale 'Under a Jack and the Beanstalk tree (a pillar of melted amber)', the latter being a childlike memory of looking high up to an unreachable imaginary world of 'castles-in-the-air'; and the amber colour relating to the embalmed insects frozen in time. The memories also hark back to our watery origins as in the 'slippery roots like fins veloured in moss' (Padel, 2010, p. 33). And in the same way as all our memories are related to one another making up the past and the present, so too in Nature's relationships as in Padel's reference to 'Creepers – and a strangler ficus – crafting their way to the light. Relationships! And inside, who knows what amphibia, mammals, reptiles?'

(Padel, 2010, 33). But here, despite its relationship with other beings, Padel's use of 'strangler ficus' gives the creeper a malign intent compared to Darwin's more gentle and benign 'twiners entwining twiners' (more like a marriage than a death knell).

Darwin and his cousin Emma were like 'twiners entwining twiners'. Their relationship could be said to be more than just romantic, but also Romantic in the sense of Emma holding all the fragmented pieces together, not just of their own personal lives but the physical and mental elements of their own natures that formed integral parts of Nature itself – their concepts and theories of the world, whether religious or scientific, the experiments at Down House and the inclusion of their children as helpers. Despite feeling inadequate, ugly, sick and knowing their religious differences, Darwin knew it was the right decision to ask her to marry him and this he did when visiting her parents' home at Maer Hall on 11 November 1838:

> He feels precarious, like facing a brass whip
> Of lightning over the sea. What is a self?
> A glass you can't climb out of. The bottom of a crater. (Padel, 2010, p. 62)

Despite his later theory of transmutation, here at least he feels that he cannot change his self. Padel gives the reader the impression that Darwin's world is almost predetermined and that his theories come from a randomly shaken box. Perhaps if he shakes the box a bit harder the right theories might pop out (could these be the ones that would satisfy Emma?):

> He studies the surface of what he can cling to
> like a wounded angel, wondering how pieces of theory
> fall at him out of a box. He keeps shaking the box. (Padel, 2010, p. 62)

Maybe he is not perfect but he is not a fallen angel, even if he is a wounded one. Either way, he is very much a part of the Nature that is swirling around him when he is preparing himself to ask Emma to marry him. It is almost as if he is a sea creature selecting, or being selected by, his mate, transporting the readers back to primordial life in the sea using 'liquid metaphors for feeling and for mind!' such as 'nature foaming between them', 'the pool of his question' and the 'floaty abyss of not being, ever being, all his life, handsome enough' (Padel, 2010, p. 62) – Darwin always having self-doubt, yet Emma having complete faith in him regardless of their differences in religious faith.

Like the *Beagle* coming out of a storm, there is a certain calm as Darwin and Emma join forces to become one, creating the feeling in Darwin of being washed open, 'an oyster cleansed of grit'. His life, his memories, his personal histories, or 'rastros', have all been rearranged by this new-life-to-be with Emma, transformed into the tropical forest he found so intoxicating in South America, Darwin becoming the 'roses burning alive' and Emma becoming 'the haze above the forest'. Emma becomes the driving force in his life like an 'engineer' but also gives Darwin the Miltonic vision of a blind man. The use of 'engineer' is an interesting turn of phrase as it conjures up the Romantic view of industrial progress depicted in Erasmus Darwin's poems as well as giving Emma an equal footing in a predominantly male world of engineers at the time of Charles Darwin. This is slightly ironic as Padel is using a word that accurately describes his feelings looking back from this point in time but which is not one he would have used due to the Victorian view of women. However, it seems natural that this poem should occur about halfway through the book, creating the impression that Emma is like a spine holding both the book and their lives together, both mentally and physically. And this happiness is expressed by the pollinating bees regenerating life through the production of honey, injecting sweetness into his life, giving him the strength and structure of a backbone enabling him to carry out his research:

> She'd be engineer of all his happiness. Bees
> shifted honey-bags up his spine. He was roses
> burning alive, and she was the haze
> above tropical forest plus the unfathomed riches
> within. Like giving to a blind man eyes. (Padel, 2010, p. 65)

The imagery Padel uses to describe Darwin's and Emma's feelings for each other and their beliefs in salvation and morality are all intertwined with organic Nature. Emma's pale face is compared to the same vaulting of the Cordillera church in which he sought illumination from reading Milton's *Paradise Lost*. It is ironic that this solid vaulting of Emma's can be seen as a solid form of stability reflecting strength of faith in God and strength of feeling for himself, yet a roof of faith he cannot accept over his own head despite it forming a foundation for their own relationship. 'He accepts Christian morality' but not its religious origin as 'He's sure, now, morality has organic origin'. This forms the thesis of his *Descent* and is subconsciously made clear to him when he sees 'a quiff of feather on her skirt' (Padel, 2010, p. 67). This captures the theory of sexual selection yet to be developed, in which humans and pheasants follow the same natural laws of mating. In a moving passage in which Padel describes Emma praying that Darwin will change his belief in salvation, their differences become entwined in a naturalistic description. Emma's thinking mind is described as a mix of chemical changes and marine life such as coral and dolphins, further emphasising the organic origins of life and the physical origins of morality:

> [...] A sudden flush
> like coral cloud in eastern sky, a photochemical exchange
> on the face he's known since he was small, marks passage
> of the prayer that must be moving in her mind
> like dolphins underwater. *God grant he change.* (Padel, 2010, p. 67)

It is also ironic that Emma's face changes as her mind prays for Darwin to change his belief in salvation, a face that did not express these strengths of feeling when she was small, and Darwin's belief moving in the opposite direction, along with his theories.

In the poem 'A Path Around a Lake' (Padel, 2010, pp. 68–70), their relationship is described shortly after they are married and their differences in faith are highlighted by references to 1 Corinthians and John 3.16 which act as a chorus, similar to the chorus in T. S. Eliot's *Murder in the Cathedral*. Although Emma knows reading the New Testament won't change Darwin's views, she would nevertheless like him to read 'Our Saviour's farewell to His disciples' as this underlines her belief that '*Your faith should stand in the power of God, not the wisdom of man*' (from 1 Corinthians 2.5).[2] For Darwin, his faith is in the belief that wisdom and morality have evolved from the power of Nature and its laws, and not from God. For Emma this

---

2   Incorrectly referred to as 1 Corinthians 2.4 in the poem.

faith 'in the power of God' is exemplified in Christ's farewell to His disciples in John 13.1–17.26 in which Jesus leaves the material world and His material body to go to the Spiritual world. For Darwin, morality came from organic Nature, but for those having faith 'in the power of God, morality came from Divine Love and this was reflected in Jesus washing the feet of his disciples, with the exception of the absent Judas. The act of cleansing was an act of Love, and the act of being cleansed was an act of Faith: 'All of you except one are clean' (John 13.10). All this toing and froing of thought between the material and the spiritual is captured in Padel's poem 'Cold Water Thrown on the Head' which covers the period in December 1838 when Darwin and Emma share their first home together in London at Great Marlborough Street. Here there is a hint of the *Rime of the Ancient Mariner* in which Darwin is almost depicted as a faithless Judas with unclean feet:

> His feet hurt. Five baby cockroaches, hatched below
> the boards of Great Marlborough street kitchen,
> gather like amber soot in a crack of skirting
> ready to whisk inside the cistern when he moves. (Padel, 2010, p. 71)

The hurting of his feet suggests tired feet requiring love and attention, but in proximity to the cockroaches which symbolize dirt, they also indicate the need to be washed, and in so doing to cleanse his spirit. It is perhaps no coincidence that the cockroaches are babies generated by Darwin's 'dirty' sacrilegious thoughts of materialism, thus gathering 'like amber soot'. However, the 'amber soot' could be seen as optimistic as it suggests fossilized soot hinting that such thoughts will not be regarded as 'dirty' or sooty in the future. This is supported by the vision of the cockroaches 'ready to whisk inside the cistern when he moves' indicating the cleansing process of the water through the movement of Darwin's ideas towards a view of Nature in which the material and the spiritual become one, in which Love and Faith are one, albeit a belief in a different kind of power. In John 16.17 Jesus tells his disciples '*In a while you shall not see me. A little while more and you shall*' (Padel, 2010, p. 71).[3] For the

---

3   The quote from John 16.17 is incorrectly referenced as John 17.16.

disciples the pain of the disappearance and the reappearance is like the pain a mother experiences at birth to be forgotten as the delight of the birth overshadows the pain (John 16.21). Padel quotes Darwin as equating this pain and pleasure as being similar to having cold water thrown over his head; the effect creates similar feelings that can be 'considered spiritual':

> [...] Argument
> for materialism: that cold water, thrown on the head,
> produces a frame of mind very similar to feelings
> considered spiritual. If, as I hold, thought and perception rise
> from physical action on the brain [...]. (Padel, 2010, p. 71)

This was an experience Darwin got used to during his cold water treatments at Malvern, also experienced by his ten-year-old daughter Annie before she died.

In 'A Natural History of Babies', Padel covers Darwins's observations in his notebooks of all ten of his babies. He pictures them as being 'like painted bison, bearded and Palaeolithic in the bowels of a cave, shaking their shaggy selves off a streaming wall, dodging stalactites, tossing dreadlocks, [...]. A *whang* of feathered forelegs over rock' (Padel, 2010, p. 84). The 'bowels of a cave' are reminiscent of the 'bowels of the earth' Erasmus Darwin described when visiting caves in the North of England, with their 'streaming walls' from dripping ground water and underground streams, and stalactites. The 'rastro' images of the cave paintings of bison take the reader back to the Palaeolithic period of about 2.6 million years ago when Homo Habilis were developing stone tools. Comparing his babies to those bison pictures creates a chain-link through the 'rastro' footprints to the beginnings of human time, forming part of Darwin's 'origins' in his 'tree of life'. This is further emphasized through Padel's use of the comparison of Darwin's babies 'in movement and expression' to 'Jenny, the orang-utang' (Padel, 2010, p. 84).

The poems blend the everyday and the sublime, or supernatural, like the Wordsworthian and Coleridgean *Lyrical Ballads*. For example, Padel quotes Darwin as saying 'I am unusually well but excitement and fatigue bring on dreadful flatulence', but then adding in her own words 'He doesn't say "fart". Maybe he doesn't know the word' (Padel, 2010, p. 108). In the poem 'Painting the Bees', Padel captures this mix within the swarm of ideas buzzing around in his head creating a dizzying effect of poetic descriptions of Nature and their effect on his physical and mental constitution. The reader is taken from:

> The questions buzz at him like birds. They cling
> like burrs, delight him like children, paw like dogs.
> They scratch, torment and swarm; they pollinate
> like wasps (Padel, 2010, p. 113),

to:

> Vomit. Panic. Vomit. Nerves
> in his neck. That wagtail on the lawn –
> what muscles make it flick like that; what purpose
> does it serve? Flatulence. Bad headache. (Padel, 2010, p. 113)

The questions about the laws of Nature become transformed into attacking creatures or plants from Nature almost as if they are rebuking him for doing so. They buzz at him, cling to him, paw at him, scratch him. Even the pollinators are wasps rather than friendly bees. And the effect of the twinging nerves in his neck are mirrored by the flicking action of the wagtail. But from the height of naturalistic poetry, the reader is brought back down to the everyday language of Darwin's sickness: 'vomit', 'panic', 'flatulence', 'headache'. The poem ends with memories of his children's play and their help with his quirky experiments:

> Horace finds him snake and lizard eggs, Frank plays bassoon to worms (are they really deaf?) and even flowers to see how they like vibration. He strings all seven children over long grass and scabious in a chain, to paint and track the bumblebees who pollinate red clover. (Padel, 2010, p. 114)

These memories of his children's childhood capture the childlike and almost innocent nature of Darwin's pursuits as if he too were a child, in an atmosphere in which the distinction between the children's scientific experiments and their games is blurred, creating a world in which scientific fantasies or 'castles-in-the air' are permissible. In his children's garden of play, Darwin is able to become a child again, and with his children is able to revisit play, stretching his and their imaginations to their limits.

In the poem 'The Black and the Green' (Padel, 2010, pp. 127–8), Padel cleverly weaves the sad funeral of Darwin and Emma's baby Mary with descriptions of Nature and deep time that mirror the event alongside reference to the joint reading of Wallace and Darwin's papers at the Linnean Society in London, read on the same day. A touch of the supernatural is reflected in the description of Emma 'facing the narrow spire like a small Dutch wizard's hat' as if she has been endowed with some hidden powers, yet perhaps not as physically solid as the 'narrow spire'. The 'hover-fly just above her shoulder on translucent wings' gives the impression of an angel taking the soul of the baby up to heaven (the afterlife being an obsession of Emma's), and their 'translucent' nature creates a feeling of intense light in contrast to Emma's blackness, creating a note of optimism. Even the grave takes on the shape of the baby as if it were somehow still alive, with 'A little mouth of opened earth by the west door'. Padel describes the waiting worms as Darwin would, as almost human, sensitive to light and, like Darwin, putting herself, through Darwin, under the skin of the worms wondering what it would be like to be disturbed:

> Worms will be wriggling away from light
> What is it like, to be disturbed in there? (Padel, 2010, p. 127)

This poem captures life and death, 'redemption after suffering', evolution both in theory in the form of Darwin's forthcoming publication of the *Origin* and through examples of deep time in the surrounding Nature. The 'redemption after suffering' is reflected in 'the infinity of loss' and his feeling of being 'ashamed', similar to the feelings of the Ancient Mariner. The 'infinity of loss' is reflected in deep time as inorganic and organic matter, forever changing, dying and being born again:

> Flints, mashed
> into shape by geothermal pressure: discoloured teeth
> in the wall. As they walk out, Bessy pulls to stroke
> the grain of an upright in the porch. An oak beam
> five hundred years, perhaps, in the making. (Padel, 2010, pp. 127–8)

The flints in the church wall are the products of deep time yet at the same time put there by Man. This relationship is mirrored in their likeness to discoloured teeth. And Bessy the maid's stroking of the five-hundred-year-old oak beam brings the past and the present together through her touch. At the end of the poem, the optimism of life breaking through the darkness of death is captured through the contrasting colours of the black of the coffin and the bright green young grass on which his children are standing. Life, as in Nature, continues through those that survive:

> He looks at his children
> on bright grass. The little coffin. Black and green. (Padel, 2010, p. 128)

The contrast of the 'black and green' is reminiscent of the description of Darwin's intoxicating visit to Bahia when he contrasts the green orchid leaves against their black branches 'like green flames over Bibles' (Padel, 2010, p. 32). Towards the end of his life as he gets older and suffers more from his various illnesses, Darwin grows a beard due to 'a bout of eczma' giving him a 'biblical' look (Padel, 2010, p. 132). Again there is a hint of the supernatural with its mix of the paradoxical and the enigmatic akin to the mystical 'Dutch wizard's hat' in mirrored opposition to the 'narrow spire'. Darwin's laws of Nature are portrayed as self generating yet there is still a mystery as to their first primordial origins. And this mystery comes through in the description of Darwin as a biblical figure, as a prophet endowed with the powers to reveal the secrets of Nature. Having distanced himself from God as a First Cause, he has somehow physically (and perhaps spiritually) evolved into a form reflecting those powers. This enigmatic paradox is emphasized in Padel's poem 'They Give Him a Coat of Fur' (Padel, 2010, p. 138). The fur coat is given to Darwin as a surprise 'to keep out cold' and is described as:

> A soft mountain,
> colour of Alaskan forest in the dusk, each hair
> a dark blaze of loam. (Padel, 2010, p. 138)

Not only does it appear as part of Nature through its comparison to 'a soft mountain' and 'colour of Alaskan forest' but it harks back to a deeper time through the description of the hairs as 'a dark blaze of loam' representing decayed vegetable matter, also reflected by the 'dusk'. When Darwin puts it on:

> He feels like a dream of the north –
> of hibernation – of the Earth's internal heat. He comes out
> to the hall, shrouded in animal, a priest
> with tears in his eyes. (Padel, 2010, p. 138)

Like an animal, he appears to come out of hibernation, but a deeper hibernation from a deeper time, 'the Earth's internal heat' suggesting he has risen up from the bowels of the earth like some strange creature from one of Erasmus Darwin's caves. Again, his appearance of being 'shrouded in animal' gives him the air of mystery, making him look like 'a priest'.

Padel's biographical collection of poems draws out all that is Romantic about Charles Darwin, in particular Darwin's 'rastro' method of tracing the footprints of Nature's origins through his ability to see the interrelatedness of Nature as an organic whole. His descriptions of creatures, and even plants, as being almost human, enable the reader to empathize with creatures who could have been our ancestors, enabling the reader to enter their minds, realising the commonality of being that all creatures share. It is therefore very fitting that towards the end of his life Darwin is seen almost to transmute into an animal through his fur coat, the loamy coloured hairs of the fur holding past secrets like the plumage of the Argus pheasant, each mirroring a moment in history. It is apt, therefore, that Darwin's great-great granddaughter, Ruth Padel, should be able to take the 'nauseating' out of poetry and use it as a way of drawing out the Romantic nature of Darwin and his work, and, in so doing, enable the reader to appreciate the beauties of Nature with its 'web of affinities' and its 'entangled bank' through the eyes of Darwin.

In the introduction to this book, reference was made to Percy Bysshe Shelley's 'A Defence of Poetry' (Shelley, 2008) in order to highlight the importance of Darwin's Romantic imagination and how this contributed to the development of his ideas. One of Darwin's key imaginative abilities was his ability to 'put himself in the place of another' making the pains and pleasures of his species his own. For Shelley, this ability is expressed through poetry (which applies to writing in general, with the exception of story writing which he regards as 'partial' and only applying 'to a definite period of time' as opposed to poetry which 'is universal') (Shelley, 2008, pp. 45–6). As expressed by Shelley:

> A man, to be greatly good, must imagine intensely and comprehensively; he must put himself in the place of another and of many other; the pains and pleasures of his species must become his own. The great instrument of moral good is the imagination; and poetry administers to the effect by acting upon the cause. (Shelley, 2008, p. 48)

For Darwin, he went beyond putting himself in the place of others of his species; he put himself in the place of *all* species enabling him to be sympathetic not only to mankind but to the whole of Nature. Darwin, through his Romantic imagination and poetic writing, was able to penetrate the veil covering the secrets of Nature, enabling the human mind to break through the familiar and start to understand and appreciate the unfamiliar: 'Poetry lifts the veil from the hidden beauty of the world and makes familiar objects be as if they were not familiar' (Shelley, 2008, p. 47).

In Darwin's final paragraph in the *Origin* in which he refers to 'the war of nature, from famine and death, the most exalted object which we are capable of conceiving, namely, the production of the higher animals, directly follows', and where a few lines later he ends the book with 'from so simple a beginning endless forms most beautiful and most wonderful have been, and are being, evolved' (Darwin 1985, pp. 459–60), the lines from Shelley come to mind written thirty-eight years earlier:

> Poetry turns all things to loveliness; it exalts the beauty of that which is most beautiful, and it adds beauty to that which is most deformed; it marries exultation and horror, grief and pleasure, eternity and change; it subdues to union under its light yoke, all irreconcilable things. It transmutes all that it touches, and every form moving within the radiance of its presence is changed by wondrous sympathy to an incarnation of the spirit which it breathes: its secret alchemy turns to potable gold the poisonous waters which flow from death through life; it strips the veil of familiarity from the world, and lays bare the naked and sleeping beauty, which is the spirit of its forms. (Shelley, 2008, p. 64)

This is the optimism captured by the poetry of Shelley, Charles Darwin and Ruth Padel, in which the Romantic imagination transforms the mix of life and death into 'loveliness' and 'beauty' by creating a union of all things irreconcilable, whether this be 'horror', 'grief', 'pleasure', 'eternity' or 'change'. In Darwin we can see all this in his 'war of nature' as well as in his personal life, whether this be in the delights of his family and research or the death of his children and friends, or indeed his own sickness. And this common experience that Darwin feels all species share is the same sympathy that comes across as 'an incarnation of the spirit', whether this be in Nature or poetry. This is the Romantic imagination generated by Darwin that enables the reader, through sympathy with Nature, to breathe in its spirit, turning its 'secret alchemy [...] to potable gold the poisonous waters which flow from death through life'. The reader of *The Origin of Species* is actively encouraged to develop this sympathy with Nature by imaginatively stepping back 'to contemplate' Nature's interrelatedness. Darwin does this by putting himself forward as a role model through his own contemplation and reflection:

> It is interesting to contemplate an entangled bank, clothed with many plants of many kinds, with birds singing on the bushes, with various insects flitting about, and with worms crawling through the damp earth, and to reflect that these elaborately constructed forms, so different from each other, and dependent on each other in so complex a manner, have all been produced by laws acting around us. (Darwin, 1985, p. 459)

In this passage, through his simple words, he is able to create a calm and tranquil view of Nature that draws in the reader making her/him a participant in the tranquillity, despite its hidden complexity. Although the bank is 'entangled' it is 'clothed' with plants rather than taken over by overgrown weeds. The bird song is not described as noisy and the worms crawl like other insects rather than wriggle, which might put off the reader. These gentle descriptions help gain the reader's sympathy and trust but at the same time without glossing over the fact that all organisms are 'elaborately constructed' and 'dependent on each other in so complex a manner'. This is the art of Charles Darwin's Romantic poetry enabling the reader to find sympathy with and optimism in Nature.

Bibliography

Primary Sources

Barrett, Paul H. et al. (2008). *Charles Darwin's Notebooks, 1836–1844; Geology, Transmutation of Species, Metaphysical Enquiries*. London: Natural History Museum and Cambridge: Cambridge University Press.
Burkhardt, Frederick (ed.) (2008). *Charles Darwin: The 'Beagle' Letters* (1831–1836). Cambridge: Cambridge University Press.
Burkhardt, Frederick (ed.) (2009a). *The Correspondence of Charles Darwin. Anniversary Set 1821–1860. Volume 1 1821–1836*. Cambridge: Cambridge University Press.
Burkhardt, Frederick (ed.) (2009b). *The Correspondence of Charles Darwin. Anniversary Set 1821–1860. Volume 2 1837–1843*. Cambridge: Cambridge University Press.
Burkhardt, Frederick (ed.) (2009c). *The Correspondence of Charles Darwin. Anniversary Set 1821–1860. Volume 3 1844–1846*. Cambridge: Cambridge University Press.
Burkhardt, Frederick (ed.) (2009d). *The Correspondence of Charles Darwin. Anniversary Set 1821–1860. Volume 4 1847–1850*. Cambridge: Cambridge University Press.
Burkhardt, Frederick (ed.) (2009e). *The Correspondence of Charles Darwin. Anniversary Set 1821–1860. Volume 5 1851–1855*. Cambridge: Cambridge University Press.
Burkhardt, Frederick (ed.) (2009f). *The Correspondence of Charles Darwin. Anniversary Set 1821–1860. Volume 6 1856–1857*. Cambridge: Cambridge University Press.
Burkhardt, Frederick (ed.) (2009g). *The Correspondence of Charles Darwin. Anniversary Set 1821–1860. Volume 7 1858–1859*. Cambridge: Cambridge University Press.
Burkhardt, Frederick (ed.) (2009h). *The Correspondence of Charles Darwin. Anniversary Set 1821–1860. Volume 8 1860*. Cambridge: Cambridge University Press.
Chancellor, Gordon and Van Wyhe, John (transcribed, edited and introduced by) (2009). *Charles Darwin's Notebooks from the Voyage of the 'Beagle'* (1831–1836). Cambridge: Cambridge University Press.
Darwin, Charles (1880). *The Power of Movement in Plants* (1880). Assisted by Francis Darwin. London: John Murray.
Darwin, Charles (1909). *The Foundations of the Origin of Species: Two Essays* (1842 and 1844), ed. Frances Darwin. Cambridge: Cambridge University Press.
Darwin, Charles (1962). *Voyage of the Beagle (1839)*. (1845, 2nd edn, edited by Leonard Engel). New York: Doubleday Anchor.

Darwin, Charles (1985). *The Origin of Species* (1859, 1st edn, edited with an introduction by J. W. Burrow). London: Penguin.

Darwin, Charles (1989). *Voyage of the Beagle* (1839, 1st edn, edited and abridged with an introduction by Janet Browne and Michael Neve). London: Penguin Classics.

Darwin, Charles (1995). 'Autobiography'. In: Darwin, Francis, *The Life of Charles Darwin*. London: Senate, pp. 5–54 (see *Secondary Sources*).

Darwin, Charles (2002). *Autobiographies* ('An autobiographical fragment' first published in 1903, and 'Recollections' first published in 1887). Edited by Michael Neve and Sharon Messenger. London: Penguin Classics.

Darwin, Charles (2003). *The Voyage of H. M. S. Beagle* (1839). (1860, 3rd edn, with an introduction by Richard Keynes). London; The Folio Society.

Darwin, Charles (2004). *The Descent of Man, and Selection in Relation to Sex* (1871). (1879, 2nd edn reprint, with an introduction by James Moore and Adrian Desmond). London: Penguin Classics.

Darwin, Charles (2008). *The Descent of Man, and Selection in Relation to Sex* (1871). (1877, 2nd edn, with an introduction by Richard Dawkins (2003)). London: Folio Society.

Darwin, Charles (2009a). *Charles Darwin's Beagle Diary (1831–1836)* (1839). Sioux Falls: Ezreads Publications LLC.

Darwin, Charles (2009b). *The Expression of the Emotions in Man and Animals* (1872). (1998, 3rd edn, with an Introduction, Afterword and Commentaries by Paul Ekman). London: Harper P.

Darwin, Charles (2009b). Charles Darwin's Beagle Diary (1831–1836) (1839). Sioux Falls: Ezreads Publications LLC.erennial.

Darwin, Charles and Wallace, Alfred Russel (2008). 'On the Tendency of Species to form Varieties; and on the Perpetuation of Varieties and Species by Natural Means of Selection. By Charles Darwin, Esq., F. B. S., F. L. S., & F. G. S., and Alfred Wallace, Esq. Communicated by Sir Charles Lyell, F. R. S., F. L. S., and J. D. Hooker, Esq., M.D., V. P. R. S., F. L. S., &c. (Read July 1st, 1858.) London, June 30th, 1858'. London: Linnean Society.

Darwin, Charles and Wallace, Alfred Russel (2008). 'On the Tendency of Species to form Varieties; and on the Perpetuation of Varieties and Species by Natural Means of Selection' (1858), *The Linnean Society of London: The Linnean Special Issue Number 9: The Survival of the Fittest: Celebrating the 150th anniversary of the Darwin-Wallace theory of evolution*, edited by Brian Gardiner, Richard Milner and Mary Morris.

Darwin Correspondence Project (2014a). 'Darwin's Reading Notebooks'. University of Cambridge. Available at: <http://www.darwinproject.ac.uk/darwins-reading-notebooks> (Accessed on 27 March 2014).

Darwin, Erasmus (2007). *The Botanic Garden Part II: The Loves of the Plants* (1791). Teddington: The Echo Library.
Darwin, Erasmus (2008). *The Temple of Nature; or; the Origin of Society: A Poem, with Philosophical Notes* (1802). Teddington: The Echo Library.
Darwin, Erasmus (2010). *The Botanic Garden Part I: The Economy of Vegetation* (1791). Whitefish, MT: Kessinger Publishing.
Keynes, Richard D. (edited and introduced by) (2009). *Charles Darwin's Beagle Diary (1831–1836)*. (Reprinted Cambridge University Press edition 1988). Sioux Falls, SD: Ezreads Publications LLC.
King-Hele, Desmond (ed.) (2003). *Charles Darwin's The Life of Erasmus Darwin* (1879). Cambridge: Cambridge University Press.

Secondary Sources

Bayliss, Bill (2012). *Fuega Basket, Captain Fitzroy, Charles Darwin and The Beagle*. Available at: <http://www.walthamstowmemories.net/pdfs/Bill%20Bayliss%20-%20Fuegia%20Basket.pdf> (Accessed on 12 March 2015).
Beer, G. (1983). *Darwin's Plots: Evolutionary Narrative in Darwin, George Eliot, and Nineteenth-Century Fiction*. Cambridge: Cambridge University Press.
Beer, G. (2010). 'Darwin and Romanticism', *The Wordsworth Circle*, vol. 41, issue 1 – Winter 2010, pp. 3–9.
Bellis, Mary, (2013). *History of Photography*. Available at: <http://www.inventors.about.com/od/pstartinventions/a/stilphotography.htm> (Accessed on 16 April 2013).
Blake, William (2015). 'The Marriage of Heaven and Hell' (1790–1793). Available at: <http://www.bartleby.com/235/253.html> (Accessed on 20 October 2015).
Bortoft, Henri (2010). *The Wholeness of Nature: Goethe's Way toward a Science of Conscious Participation in Nature*. Edinburgh: Floris Books.
Botting, Douglas (1973). *Humboldt and the Cosmos*. London: Sphere Books.
Bowie, Andrew (2003). *Aesthetics and Subjectivity: from Kant to Nietzsche*. Manchester and New York: Manchester University Press.
Bowler, Peter J. (1989). *Evolution: The History of an Idea*. London: University of California Press.
Brooke, John Hedley (2003). 'Darwin and Victorian Christianity'. In: Jonathan Hodge and Gregory Radick (eds), *The Cambridge Companion to Darwin*. Cambridge: Cambridge University Press.

Brooklyn College (2011). *The Angel in The House*. The City University of New York. Available at: <http://www.academic.brooklyn.cuny.edu/english/melani/novel_19c/thackeray/angel.html> (Accessed on 12 June 2014).

Broughton, Trev Lynn (1999). *Men of letters, writing lives: masculinity and literary auto/biography in the late Victorian period*. London: Routledge.

Browne, Janet (2003a). *Charles Darwin: Voyaging*. London: Pimlico.

Browne, Janet (2003b). *Charles Darwin: The Power of Place*. London: Pimlico.

Buckland, Raymond (2005). *The Spirit Book: The Encyclopedia of Clairvoyance, Channeling, and Spirit Communication*. Canton, MI: Visible Ink Press.

Bumblebee Conservation Trust (2013). *Bumblebee Nests*. Available at: <http://www.bumblebeeconservation.org/about-bees/faqs/bumblebee-nests/> (Accessed on 4 April 2013).

Burke, Edmund (2010). *Reflections on the French Revolution* (1790). Available at: <http://www.socserv.mcmaster.ca/econ/ugcm/3113/burke/revfrance.pdf> (Accessed on 8 November 2010).

Burns, J. H. (2005). 'Happiness and utility: Jeremy Bentham's Equation'. In: *Utilitas*, 17, pp. 46–61.

Burrow, J. W. (1985). 'Note on this Edition'. In: Darwin, Charles, *The Origin of Species* (1859). London: Penguin.

Chambers, Robert (2010). *Vestiges of the Natural History of Creation* (1844). Memphis, Tennessee: General Books.

Clark, J. W. and Hughes, T. M. (1890). *The Life and Letters of Adam Sedgwick*, 2 vols. Cambridge: Cambridge University Press.

Coleridge, Samuel Taylor (2013). *The Eolian Harp* (1795). Poetry Foundation. Available at: <http://www.poetryfoundation.org/poem/183957/> (Accessed on 5 November 2013).

Conrad, Joseph (1976). *Heart of Darkness (1899)* and *Typhoon*. London: Pan Books.

Coplestone, Frederick (1967). *A History of Philosophy Volume VIII. Modern Philosophy: Bentham to Russell. Part I. British Empiricism and the Idealist Movement in Great Britain*. New York: Image Books.

Darwin, Francis (1995). *The Life of Charles Darwin*. London: Senate.

Davidson, Paul G. (2012). *Evolution is the Core of Modern Taxonomy (Systematics)*. Available at: <http://www.buildingthepride.com/faculty/pgdavison/BI%20101/evolution_is_the_core_of_modern_.htm> (Accessed on 6 December 2012).

Desmond, Adrian and Moore, James (2004). 'Introduction'. In: Darwin, Charles (2004) *The Descent of Man, and Selection in Relation to Sex* (1879, 2nd edn). London: Penguin Classics.

Desmond, Adrian and Moore, James (2009). *Darwin*. London: Penguin Books.

Desmond, Adrian and Moore, James (2010). *Darwin's Sacred Cause: Race, Slavery and the Quest for Human Origins*. London: Penguin Books.
Dunkerton, J. and Howard, H. (2009). 'Sebastiano del Piombo's "Raising of Lazarus": A History of Change'. *National Gallery Technical Bulletin*, vol. 30, pp. 26–51.
Ellis, David (2011). *Byron in Geneva: That Summer of 1816*. Liverpool: Liverpool University Press.
Encyclopaedia Britannica (2014). 'Longinus'. Available at: <http://www.britannica.com/EBchecked/topic/347517/Longinus> (Accessed on 31 March 2014).
Encyclopaedia Britannica (2015). 'Alexander von Humboldt'. Available at: <http://www.britannica.com/biography/Alexander-von-Humboldt> (Accessed on 22 September 2015).
Encyclopaedia Britannica (2015b). 'Biogenetic Law'. Available at: <http://www.britannica.com/science/biogenetic-law> (Accessed on 23 October 2015).
European Graduate School, The (2015). 'Ernst Haeckel – Biography'. Available at: <http://www.egs.edu/library/ernst-haeckel/biography/> (Accessed on 23 October 2015).
Fara, Patricia (2003). *Sex, Botany and Empire*. Cambridge: Icon Books UK.
Fara, Patrica (2012). *Erasmus Darwin: Sex, Science and Serendipity*. Oxford: Oxford University Press.
Ferguson, Alfred P. (ed.) (1964). *The Journals and Miscellaneous Notebooks of Ralph Waldo Emerson*. Cambridge, MA.: Harvard University Press.
Gaul, Marilyn (1979). 'From Wordsworth to Darwin: "On to the Fields of Praise"', *The Wordsworth Circle*, vol. 10, issue 1 – Winter 1979, pp. 33–48.
Goethe, Johann Wolfgang von (1949). 'The Metamorphosis of Plants'. In: Rudolf Magnus (1949), *Goethe as a Scientist*, translated by Heinz Norden. New York: Henry Schuman.
Goethe, Johann Wolfgang von (1970). *Italian Journey* (1786–1788). Translated by W. H. Auden and Elizabeth Mayer. London: Penguin.
Goethe, Johann Wolfgang von (1985–1998). 'Die Faultiere und die Dickhäutigen', *Zur Morphologie*, in *Sämtliche Werke nach Epochen seines Schaffens* (Münchner Ausgabe). Edited by Karl Richter et al. 21 vols. Munich: Carl Hanser Verlag.
Goethe, Johann Wolfgang von (1985–1998b). *Erster Entwurf einer allgemeinen Einleitung in die vergleichende Anatmie, ausgehend von der Osteologie*, in *Sämtliche Werke nach Epochen seines Schaffens* (Münchner Ausgabe). Edited by Karl Richter et al. 21 vols. Munich: Carl Hanser Verlag.
Goethe, Johann Wolfgang von (1985–1998c). *Tag- und Jahres-Hefte* (1790). In: *Sämtliche Werke nach Epochen seines Schaffens* (Münchner Ausgabe). Edited by Karl Richter et al. 21 vols. Munich: Carl Hanser Verlag.

Goethe, Johann Wolfgang von (1988). 'Studies for a Physiology of Plants'. In: *Scientific Studies*, edited and translated by Douglas Miller. New York: Suhrkamp, 1988.
Goethe, Johann Wolfgang von (1998). *Goethes Briefe* (Hamburger Ausgabe). Edited by Karl Robert Mandelkow, 4th edn, 4 vols. Munich: C. H. Beck.
Goethe, Johann Wolfgang von (2009). *The Metamorphosis of Plants* (1790). Introduction and photography by Gordon L. Miller. London: The MIT Press.
Hadot, Pierre (2006). *The Veil of Isis*. London: The Belknap Press of Harvard University Press.
Hall, T. D. (1995). 'Influence of Malthus and Darwin on the European Elite'. Available at: <http://www.trufax.org/avoid/manifold.html> (Accessed on 25 October 2010).
Halmi, N. (2012). 'Coleridge's Ecumenical Spinoza'. Available at: <https://www.erudit.org/revue/ravon/2012/v/n61/1018604ar.html> (Accessed on 20 October 2015).
Healey, Edna (2002). *Emma Darwin: The Inspirational Wife*. London: Review/Headline Book Publishing.
Heffer, Simon (2013). *High Minds: The Victorians and the Birth of Modern Britain*. London: Random House.
Herbert, Sandra and Barrett, Paul H. (2008). '*Notebook M* Introduction'. In: Barrett, Paul H. et al. (2008) *Charles Darwin's Notebooks, 1836–1844; Geology, Transmutation of Species, Metaphysical Enquiries*. London: Natural History Museum and Cambridge: Cambridge University Press.
Herbert, Sandra and Kohn, David (2008). 'Introduction'. In: Barrett, Paul H. et al. (2008) *Charles Darwin's Notebooks, 1836–1844; Geology, Transmutation of Species, Metaphysical Enquiries*. London: Natural History Museum and Cambridge: Cambridge University Press.
Hodge, J. (2003). 'The notebook programmes and projects of Darwin's London years'. In: Jonathan Hodge and Gregory Radick (eds), *The Cambridge Companion to Darwin*. Cambridge: Cambridge University Press, pp. 40–68.
Holmes, Richard (2009). *The Age of Wonder*. London: Harper Press.
Hulme, T. E. Hulme (1911). 'Romanticism and Classicism'. Available at: <http://www.poetryfoundation.org/learning/essay/238694> (Accessed on 17 Nov 2011).
Humboldt, Alexander von (1806). *Beobachtungen aus der Zoologie und vergleichenden Anatomie: Gesammelt auf einer Reise nach den Tropen-Ländern des neuen Kintinents*. Tübingen: F. G. Cotta.
Humboldt, Alexander von (1849–1870). *Cosmos: A Sketch of a Physical Description of the Universe*. Translated by E. C. Otté, 5 vols (1845–1861). London: Henry G. Bohn.
Humboldt, Alexander von (1995). *Personal Narrative* (1799–1804). Abridged and Translated by Jason Wilson. London: Penguin.

Humboldt, Alexander von (1997). *Cosmos: A Sketch of a Physical Description of the Universe* Volume 1 (1845). Translated by E. C. Otté; Introduction by Nicolaas A. Rupke. Baltimore and London: The John Hopkins University Press.

Hurley, Chris (2009). 'Charles Darwin and Evolution 1809–2009'. Christ's College, Cambridge. Available at: <http://www.darwin200.christs.cam.ac.uk/pages/index.php?page_id=f3> (Accessed on 17 June 2014).

Huxley, T. E. (2004). 'Note on the resemblances and differences in the structure and the development of the brain in man and apes',. In: Darwin, Charles *The Descent of Man, and Selection in Relation to Sex* (1879, 2nd edn, pp. 230–40). London: Penguin Classics.

Irmscher, Christoph (2004). 'Darwin's Beard'. In: *Old Age and Ageing in British and American Culture and Literature*, edited by Christa Jansohn, pp. 87–105. Muenster: LIT Verlag.

Jimack, P. D. (1993). 'Introduction'. In : Rousseau, Jean-Jacques *Emile or on Education* (1762). London: Everyman, J. M. Dent.

Jones, Steve (2009). *Darwin's Island: The Galapagos in the Garden of England*. London: Little, Brown.

Kant, Immanuel (1836). *Metaphysics of Ethics*, translated by J. W. Semple. Edinburgh: Thomas Clark.

Kant, Immanuel (1889). *Critique of Practical Reason* (1788). Translated by Thomas Kingsmill Abbot, 4th edn. London: Longman.

Kant, Immanuel (1914). *Critique of Judgement* (1790). Translated by J. H. Bernard (1892), 2nd revised edn. Available at: <http://www.oll.libertyfund.org/titles/1217> (Accessed on 22 October 2015).

Kant, Immanuel (1957). *Prolegomena zu einer jeden künftigen Metaphysik dies als Wissenschaft wird auttreten können* (1783). In: *Immanuel Kant Werke*. Edited by Wilhelm Weischedel, 6 vols. Wiesbaden: Insel Verlag.

Keynes, R. D. (ed.) (1988). *Charles Darwin's Beagle Diary*. Cambridge: Cambridge University Press.

King-Hele, Desmond (ed.) (1981). *The Letters of Erasmus Darwin*. Cambridge: Cambridge University Press.

King-Hele, Desmond (1999). *Erasmus Darwin: A Life of Unequaled Achievement*. London: Giles de la Mare Publishers Limited.

King-Hele, Desmond (ed.) (2003). *Charles Darwin's The Life of Erasmus Darwin*. Cambridge: Cambridge University Press.

Kreis, Steven (2009a). 'The History Guide: Lectures on Modern European Intellectual History. Lecture 16: The Romantic Era'. Available at: <http://www.historyguide.org/intellect/lecture16a.html> (Accessed on 8 December 2010).

Kreis, Steven (2009b). 'The History Guide: Lectures on Modern European Intellectual History. Lecture 9: Écrasez l'infâme!: The Triumph of Science and the Heavenly City of the 18th Century Philosophe'. Available at: <http://www.historyguide.org/intellect/lecture9a.html> (Accessed on 8 December 2010).

Lack, Walter H. (2009). *Alexander von Humboldt and the Botanical Exploration of the Americas*. London: Prestel Publishing.

Landow, G. P. (2013). 'Evolutionary Theory before Darwin'. *The Victorian Web*. Available at: <http://www.victorianweb.org/science/darwin/darwin2.html> (Accessed on 19 November 2013).

Larrimore, Mark (2008). 'Antinomies of race: diversity and destiny in Kant', *Patterns of Prejudice*, vol. 42, nos 4–5, pp. 341–63.

Levine, George (2008). *Darwin Loves You*. Princeton and Oxford: Princeton University Press.

Levine, George (2011). *Darwin the Writer*. Oxford: Oxford University Press.

Lyell, Charles (1977). *Principles of Geology* (1830–1833). Edited by James A. Secord. London: Penguin Books.

Lyons, Eric and Butt, Kyle (2013). 'Darwin, Evolution and Racism'. Apologetics Press. Available at: <http://www.apologeticspress.org/apcontent.aspx?category=9&article=2654/> (Accessed on 12 November 2013).

Mackintosh, James (1837, 2nd edn). *Dissertation on the Progress of Ethical Philosophy, Chiefly During the Seventeenth and Eighteenth Centuries*. Edinburgh: Adam and Charles Black.

Malthus, Thomas R. (2008). *An Essay on the Principle of Population* (1798). Oxford: Oxford University Press.

Man, Paul de (1993). 'Time and History in Wordsworth', reprinted from *Diacritics*, 17/4 (Winter, 1987), 4–17, in Cynthia Chase (ed.), *Romanticism*, Longman Critical Readers. London: Longman.

Markby, Thomas (1863). *In the Two books of Francis Bacon: Of the Proficience and Advancement of Learning, Divine and Human* (1605). London: J. W. Parker & Son.

McCrory, Donald (2010). *Nature's Interpreter: The Life and Times of Alexander von Humboldt*. Cambridge: The Lutterworth Press.

Melani, Lilia (2009). 'Romanticism'. English Department, Brooklyn College. Available at: <http://www.academic.brooklyn.cuny.edu/english/melani/cs6/rom.html> (Accessed on 9 December 2010).

Miller, Gordon L., (2009). 'Introduction'. In: *Goethe, Johann Wolfgang von (2009) The Metamorphosis of Plants* (1790). Introduction and photography by Gordon L. Miller. London: The MIT Press.

Montagu, Ashley (1971). 'Foreward' to Townsend, Joseph (1971) *Dissertation on the Poor Laws: By a Well-wisher to Mankind* (1788). London: University of California Press.

Moorman, Mary (1965). *William Wordsworth: The Later Years 1803–1850*. Oxford: Oxford University Press.
Myers, F. W. H. (1899). *Wordsworth* (English Men of Letters series edited by John Morley). London: Macmillan and Co., Limited.
National Gallery (2014). *The Raising of Lazarus*. Available at: <http://www.nationalgallery.org.uk/paintings/sebastiano-del-piombo-the-raising-of-lazarus> (Accessed on 13 May 2014).
Nichols, Ashton (ed.) (2004). *Romantic Natural Histories*. Boston & New York: Houghton Mifflin Company.
Nicolsen, Malcolm (1995). 'Historical Introduction'. In: Humboldt, Alexander von, abridged and translated by Jason Wilson. *Personal Narrative (1799–1804)*. London: Penguin.
O'Hear, Anthony (1999). *Beyond Evolution: Human Nature and the Limits of Evolutionary Explanation*. Oxford: Oxford University Press.
O'Neill, Michael (1997). *Romanticism and the Self-Conscious Poem*. Oxford: Clarendon Press.
Owen, Richard (1847). *Report on the Archetype and Homologies of the Vertebrate Skeleton, in Report of the Sixteenth Meeting of the British Association for the Advancement of Science; held at Southampton in September 1846*. London: Murray.
Owen, Richard (1849). *On the Nature of limbs, a Discourse delivered on Friday, February 9, at an Evening Meeting of the Royal Institution of Great Britain*. London: John Van Voorst.
Padel, Ruth (2010). *Darwin: A Life in Poems*. London: Vintage Books.
Page, Michael R. (2012). *The Literary Imagination from Erasmus Darwin to H. G. Wells: Science, Evolution, and Ecology*. Farnham, Surrey: Ashgate Publishing Ltd.
Paley, William (2005). *Natural Theology or, Evidences of the Existence and Attributes of the Deity* (1802). Landisville, Pennsylvania: Coachwhip Publications.
Patmore, Coventry (2012). *The Angel in the House* (1854), PoemHunter.com – The World's poetry Archive. PP 52–145. Available at: <http://www.poemhunter.com/i/ebooks/pdf/coventry_patmore_2012_3.pdf> (Accessed on 12 June 2014).
Phillips, Adam (1999). *Darwin's Worms*. London: Faber and Faber.
Plato (1971). *The Republic* (translated by H. D. P. Lee). London: Penguin Books.
Priestman, Martin (2013). *The Poetry of Erasmus Darwin: Enlightened Spaces, Romantic Times*. Farnham: Ashgate Publishing Limited.
Qvortrup, Mads (2003). *The Political Philosophy of Jean-Jacques Rousseau: The Impossibility of Reason*. Manchester: Manchester University Press.
Radick, Gregory (2003). 'Is the theory of natural selection independent of its history?'. In: Hodge, J. and Radick, G. (eds), *The Cambridge Companion to Darwin*. Cambridge: Cambridge University Press, pp. 143–167.

Richards, J. Robert (1989). *Darwin and the Emergence of Evolutionary Theories of Mind and Behaviour*. Chicago: University of Chicago Press.

Richards, J. Robert (2000). 'Kant and Blumenbach on the Bildungstrieb: A Historical Misunderstanding', *Studies in History and Philosophy of Science Part C: Studies in History and Philosophy of Biological and Biomedical Sciences*, vol. 31, no. 1, pp. 11–32.

Richards, Robert J. (2002). *The Romantic Conception of Life: Science and Philosophy in the Age of Goethe*. Chicago and London: The University of Chicago Press.

Rogers, Kara (2009). 'Beyond Darwin: Eugenics, Social Darwinism, and the Social Theory of the Natural Selection of Humans'. *Encyclopaedia Britannica Blog*. Available at: <http://www.britannica.com/blogs/2009/02/beyond-darwin-eugenics-social-darwinism-and-the-social-theory-of-the-natural-selection-of-humans/> (Accessed on 24 October 2014).

Rousseau, Jean-Jacques (1993). *Emile or on Education* (1762). London: Everyman, J. M. Dent.

Rousseau, Jean-Jacques (2008). *The Social Contract* (1762). Oxford: Oxford University Press.

Rupke, Nicolaas A. (1997). 'Introduction'. In: Humboldt, Alexander von, *Cosmos: A Sketch of a Physical Description of the Universe Volume 1* (1845). Translated by E. C. Otté. Baltimore and London: The John Hopkins University Press.

Ruse, Michael (2009). *Monad to Man: The Concept of Progress in Evolutionary Biology*. London: Harvard University Press.

Sachs, Aaron (2007). *The Humboldt Current*. Oxford: Oxford University Press.

Schelling, Friedrich (1927). *System des transscendentalen Idealismus*. In: *Schellings Werke*, Edited by Manfred Schröter, 3rd edn, 12 vols. Munich: C. H. Beck.

Schelling, Friedrich (1927b). *Einleitung zu Entwurf eines Systems der Naturphilosophie*. In: *Schellings Werk*, Edited by Manfred Schröter, 3rd edn, 12 vols. Munich: C. H. Beck.

Scofield, Martin (2003). 'Introduction'. In: Wordsworth, William and Coleridge, Samuel Taylor *Lyrical Ballads and Other Poems* (1801/1802).Wordsworth Poetry Library. Ware, Hertfordshire: Wordsworth Editions Limited.

Secord, James A. (2000). *Victorian Sensation: The Extraordinary Publication, Reception, and Secret Authorship of Vestiges of the Natural History of Creation*. Chicago and London: The University of Chicago Press.

Shelley, Mary Wollstonecraft (2015). *Frankenstein or The Modern Prometheus* (1831, 3rd edn). London: The Folio Society.

Shelley, Percy Bysshe (2008). 'A Defence of Poetry, Part 1'. *Percy Bysshe Shelley: A Defence of Poetry and other Essays* (1821), pp. 40–68. Available at: <http://www.biblioteca.org.ar/libros/167749.pdf> (Accessed on 30 May 2017).

Smith, Adam (2006). *The Theory of Moral Sentiments* (1790, 6th edn). São Paulo: MetaLibri.
Smiles, Samuel (1860). *Self-Help* (1859). London: John Murray.
Smith, Norman Kemp (1970). *Immanuel Kant's Critique of Pure Reason*. Translated by Norman Kemp Smith, 2nd edn. London: Macmillan.
Stewart, Dugald (1829). *Philosophical Essays Vol IV Part 2: Essay Second: On the Sublime*. Cambridge: Hilliard and Brown.
Stott, Rebecca (2003). *Darwin and the Barnacle*. London: Faber and Faber.
Stott, Rebecca (2012). *Darwin's Ghosts: In search of the First Evolutionists*. London: Bloomsbury
Sweet, Matthew (2001). *Inventing the Victorians*. London: Faber and Faber.
Tennyson, Alfred Lord (2006). *Selected Poems*. Selection, Introduction and Notes by Ruth Padel. London: The Folio Society.
Thomas, Dylan (1954). *Under Milk Wood*. London: J. M. Dent and Sons Ltd.
Townsend, Joseph (1971). *Dissertation on the Poor Laws: By a Well-wisher to Mankind* (1788). London: University of California Press.
Uglow, Jenny (2003). *The Lunar Men: The Friends who made the Future 1730–1810*. London: Faber and Faber.
Vandenberg, Phyllis and DeHart, Abigail (2014). 'Frances Hutcheson (1748–1832)'. *Internet Encyclopedia of Philosophy*. Available at: <http://www.iep.utm.edu/hutcheso/> (Accessed on 17 June 2014).
Wallace, Alfred Russel (1858). 'On the Tendency of Varieties to Depart Indefinitely from the Original Type' (1858). Available at: <http://darwin-online.org.uk/converted/published/1858_species_F350.html> (Accessed on 7 February 2011).
Welpley, James Davenport (1946). 'Humboldt's Cosmos'. In: *American Review: A Whig Journal of Politics, Literature, Art, and Science*, June, 3: 598–610.
Williams, Raymond (1983). *Keywords: A vocabulary of culture and society*. New York: Oxford University Press.
Wilson, A. N. (2003). *The Victorians*. London: Arrow Books.
Wilson, Jason (1995). 'Introduction'. In: Humboldt, Alexander von, Personal Narrative (1799–1804). London: Penguin.
Winkler, R. O. C. (1975). 'Wordsworth's Poetry'. In: *The Pelican Guide to English Literature: Volume 5 From Blake to Byron*. Harmondsworth: Penguin Books.
Wohl, Anthony S. (1990). 'Racism and Anti-Irish Prejudice in Victorian England'. Available at: <http://www.victorianweb.org/history/race/Racism.html> (Accessed on 10 September 2014).
Wordsworth, William (1949). *Poetical Works. The Excursion* (1814) vol. V. Edited by Ernest De Selincourt and Helen Darbishire. Oxford: Clarendon Press.

Wordsworth, William (1970). *The Prelude, or Growth of a Poet's Mind* (Text of 1805). Edited by Ernest De Selincourt and updated by Stephen Gill. Oxford: Oxford University Press.

Wordsworth, William (2015). 'Ode to Duty' (1805). Available at: <http://www.online-literature.com/wordsworth/524> (Accessed on 5 March 2015).

Wordsworth, William (2015b). 'Lines Composed a Few Miles above Tintern Abbey, on Revisiting the Banks of the Wye during a Tour. July 13, 1798'. Available at: <http://poetry.about.com/od/poems/l/blwordsworthtinternabbey.html> (Accessed on 30 April 2015).

Wordsworth, William (2015c). 'Ode: Intimations of Immortality from Recollections of Early Childhood' (1807). Available at: <http://www.poetryfoundation.org/poem/174805> (Accessed on 29 September 2015).

Wordsworth, William and Coleridge, Samuel Taylor (2003). *Lyrical Ballads and Other Poems* (1801/1802). Introduction and Notes by Martin Scofield. Wordsworth Poetry Library. Ware, Hertfordshire: Wordsworth Editions Limited.

Wulf, Andrea (2015). *The Invention of Nature: The Adventures of Alexander von Humboldt, the Lost Hero of Science*. London: John Murray.

Wyhe, John van (2014). 'The Bridgewater Treatises On the Power, Wisdom and Goodness of God As Manifested in the Creation'. Available at: <http://www.victorianweb.org/science/bridgewater.html> (Accessed on 25 July 2014).

Young, Robert M. (1985). *Darwin's Metaphor: Nature's Place in Victorian Culture*. Cambridge: Cambridge University Press.

Zunjic, Robert (2014). 'Jeremy Bentham: An introduction to the *Principles and Morals of Legislation* (1789)'. Available at: <http://www.uri.edu/personal/szunjic/philos/util.html> (Accessed on 16 June 2014).

Index

Figures are indicated by page numbers in italic print.

abolitionists 231–2
aesthetic judgement, and
    Romanticism 1–3, 72
aesthetic method 7–11, 16, 21, 36, 47,
    49n50, 51, 63, 70, 73, 77, 83, 88,
    112, 116, 120, 130, 136, 184–6, 200
aesthetic sense, and sexual
    selection 213–15
Agassiz, Louis 161
Albert, Prince 11–12
Anti-Slavery Society 55n7
archetypes 4–5, 59, 61n18, 76, 77, 79
    Goethe's concept of 81–2, 99–104,
    225
Argus pheasant, example of sexual
    selection 156–60, 225
Arkwright, Richard 226, 231

Bacon, Francis 148
Bahia Blanca, Argentina 194n3, 195, 201,
    242, 251
Banks, Joseph 236
Barlow, Nora 239
Bates, Henry Walter 67
Baudin, Nicolas 30
Beer, Gillian 5, 6, 70, 71, 72, 87, 129–30,
    199, 214–15
Bentham, Jeremy 122n3, 123n7
Berkeley, George 129
Berlin Academy of Sciences 29
'Big Bang' theory 234
Boer Wars 181
Bougainville, Louis-Antoine de 30

Boulton, Matthew 221, 228
'Bridgewater Treatises' 148n47, 167
Bristol Festival of Ideas 240
Brodie, Benjamin 180n22
Browne, Janet 177
Brunel, Isambard Kingdom 229
Burke, Edmund 64, 88
Byron, Lord 222, 223, 224

Cambridge University, introduction of
    natural sciences 12
Cambridge University Library 240
Carlyle, Thomas 166, 176n18
    French Revolution 176
Carlyon, Clement 75n11
'castles in the air' 10, 131, 133, 135, 139, 188,
    192, 243, 249
Chalmers, Rev Thomas 167
Chambers, Robert, *Vestiges of the Natural
    History of Creation* (1844) 12,
    71n5, 167, 172–6
children, status of 171
Chimborazo, Mount 53, 56n10
Coleridge, Samuel Taylor 67, 75n11, 129
    The Eolian Harp 46
    Lyrical Ballads 145, 240n1,
    243, 248
    The Rime of the Ancient Mariner 17,
    240–2
common consciousness 10, 202–3, 209,
    212, 219, 232
'community of descent' 33, 108, 115–16,
    162

'Community Selection', Darwin's theory of 2
compensation, laws of 60, 60n16, 61, 100, 103
Condorcet, Nicolas de 167
Conrad, Joseph, *Heart of Darkness* 186
consciousness 82, 120, 122n4, 173, 192, 194, 202n8, 204, 208n16
see also common consciousness; 'double consciousness'; self-consciousness
Contagious Diseases Act (1864) 171
Cook, Captain James 20, 21n8, 30, 236
Creationism
  Darwin's rejection of 30–3, 35–7, 39, 45–6, 48–9, 73–4
  Richard Owen on 35n38

Darwin, Annie 187n30, 206, 248
Darwin, Caroline 13
Darwin, Catherine 133
Darwin, Charles
  agnosticism 2, 187, 239
  anthropomorphism of 211
  and archetypes 109–15
  belief in a common humanity 14–15, 50, 55–8
  description of bees' hives 193, 232
  description of spider 203–5, 232
  and divine creation 30–3, 35–7, 39, 45–6, 48–9
  and 'double movement of prose' 191 193–6, 208n15, 219
  experiments of 59, 95, 116n23, 210n18
  guilt, sense of 241
  and habitual action 124–8, 136–7
  on Humboldt's influence 26, 68–9
  ill-health 134, 171, 172, 241
  interest in worms 196n5, 197, 202, 212, 228
  on life's origins in filth 210, 224
  on Lyell's *Principles of Geology* 92n12, 95
  and moral development 119–21, 133, 138–9, 143–8
  and 'moral reflection' 181–2
  on mutual relationships between species 27
  opposition to slavery 55–8, 122–3, 152
  and poetry 130–3, 135, 143–4, 191–2, 199–208, 218–19
  and racism 15, 55n8, 57n13, 180–1
  and Romantic materialism 137–8, 148, 162–3, 188
  use of 'Genetic Method' 113
  use of metaphors 5–7
  Victorian values of 165–6, 177–87, 219
  on women 15, 55n8, 167n6, 169–71, 212–15
Darwin, Charles, works of:
  Autobiography 167n6, 199, 212, 239
  B. Blanca notebook 201
  Coquimbo notebook 199
  Descent of Man, The (1871) 10, 13, 14–15, 55, 71, 119, 126, 132, 136, 146, 152, 182, 187, 203, 207, 213, 234
  Despoblado notebook 201
  Journal of Researches 203, 210
  Movement in Plants 202n8
  Notebook C 136
  Notebook M 126, 135, 137–8
  Notebook N 9, 126, 129–30
  Old & Useless Notes 126, 138–9, 240
  On the Origin of Species (1859) 6, 13, 32, 42, 65, 67n2, 71, 88, 108, 114, 117, 132, 174, 187, 210n18, 211, 253–4
  Rio notebook 202
  Transmutation Notebook D 6
  Voyage of the Beagle, The (1839) 7, 13, 76, 191

Index

Darwin, Emma (*née* Wedgwood) 170–1, 182, 206n11, 244–7, 250
   religious faith of 239, 241
Darwin, Erasmus
   and 'broth of chaos' 221, 227, 234
   and Galvanism 53
   and the Goddess of Nature 236–7
   and industrialization 226–8
   and influence on Darwin 10, 232–7
   and 'quasi-sexual energy' 235–7
   and the slave trade 230–2
   and sympathy 232–4
   and transmigration of spirits 228
   and vitalistic conception of materiality 221–3, 225–7
Darwin, Erasmus, works of:
   The Economy of Vegetation 224–5, 227
   The Families of Plants 235
   The Loves of the Plants 226, 235–6
   A System of Vegetables 235
   The Temple of Nature 222, 225–6, 232, 236
Darwin, George 15n19
Darwin, Robert Waring 137n33, 165, 187n32, 206n11
Darwin, Susan 133
Dawson, John William 93
De Man, Paul 132n27
death, as part of the creation process 17, 41–2, 132, 187, 212, 224–6, 228, 250–1, 253–4
deforestation, Humboldt's demonstration of 27
Descartes, René 120
Desmond, Adrian (with James Moore), *Darwin* (1992) 1, 154, 161
Diana of Ephesus 215, *216*, 236
Dickens, Charles 172
'Divergence of Character' 35, 42
diversity 6n13, 11, 20, 24, 32, 34, 40, 43, 49–50, 76, 85–7, 105, 149
Divorce Bill (1855) 171
DNA technology 32n32
'double consciousness', Darwin's concept of 9, 129–31, 188, 208n16
Down House 59, 116n23, 165n1, 186–7, 210n18

E conchis omnia (motto) 222n1
earthquakes
   at Concepcion 196
   Humboldt's study of 24
ecology, modern 21n7, 28, 58
electricity, discovery of 51
Emerson, Ralph Waldo 44
emotions
   and aesthetic judgement 3, 7, 69–70, 139, 144
   and moral reflection 181–2, 186, 232
Enlightenment, Age of 2n3
'entangled bank' metaphor 6–7, 42, 59, 71, 106, 254
eugenics 15n19, 170, 182
evolution 14, 32, 42n41, 66, 78, 156, 169, 176n18, 192, 224
   see also Oxford evolution debate (1860)
evolution, convergent 94n13
exploration, and colonial expansion 21
extinction 34, 37–43, 48–50, 66–7, 90–1, 94, 108, 127n14, 209, 225
eye, intricacy of, as argument for and against Creationism 73–4, 207
Eyre, Edward 181

Fara, Patricia 231
Fitzroy, Captain 168, 214
Fitzroy, Robert 180n22
Forster, George 20, 86
Forster, Johann 20

free will, illusion of 138
French Revolution 64
Fuegians, brought to England on the
    Beagle 168, 214

Galapagos Archiplago 94
Galton, Francis 15n19, 169–70, 179n21
Galvani, Luigi/ Galvanism 51, 53, 223
genealogical development 4–5, 35, 38,
    61, 64, 76–9, 99, 105, 108, 109,
    115–16, 119
geology, and Darwin's theory of natural
    selection 88–93, 135n31
German Romantic Movement 3, 5
God (Christian)
    and creation of individual
        species 33n33, 35, 36–7, 46,
        48–9, 63
    Darwin merges with Nature 2–3, 5,
        30, 49n49, 75, 241–2
    Darwin's loss of belief in 187
    Emma Darwin's faith in 245–7
    and Erasmus Darwin 234–5, 236
    and laws of creation 32n31
    and Man's morality 71, 123n8, 128,
        172, 174
    Paley's analogy of a watchmaker 31
    and 'providential evolution' 35n38
    'unnecessary' as cause of creation 30,
        116
goddess of nature 215, *216*, 221–2, 235,
    236–7
Godwin, William 167
Goethe, Johann Wolfgang 68, 72
    archetypes 81–2, 88, 111, 143, 198
    'Genetic Method' 5, 64, 105–6, 146
    The Metamorphsis of Plants 103, 104
Gravelet, Jean François (Blondin) 168
Great Exhibition (1851) 12
'greatest-happiness principle' 122n3
Greg, William 178

habitual action, Darwin's theories
    on 124–8
Haggard, Rider 181
Harvey, William, and circulation of the
    blood 66, 227, 230
Henslow, John Stevens 4, 25–6
Henty, Alfred 181
Herschel, John 29, 68
Herschel, William 224
Hooker, Joseph 30, 67n2, 165, 180n22,
    241
Hulme, T. E. 72
Humboldt, Alexander von
    belief in a common humanity 55–8
    influence on Darwin 5, 8, 10–11
    influences on 19–21, 26–7
    and interaction of Mind with
        Nature 83–8, 90
    organic view of nature 19–30
    physical experiments of 51–4, 229
    plant geography of 95–8
    and the sublime 75
Humboldt, Alexander von, works of:
    Cosmos 28, 29, 43–4, 56, 58,
        87, 88
    Florae Fribergensis specimen
        (1871) 20n4
    Personal Narrative 13, 21, 25, 53, 63,
        68, 76, 90
Humboldt, Wilhelm von 56
'Humboldtian Method' 8, 10–11, 21, 23,
    88
Hume, David 129
Hutcheson, Francis 123
Hutton, James, steady-state
    theory 192–3, 194n3
Huxley, Thomas 180n22

ice age, Darwin's theories on 95
imagination, Romantic concept of 8–10,
    67, 73–4, 252–4

Index

Indian mutiny (1857) 181
Infant Life Protection Bill (1871) 171
Infants and Child Custody Act
    (1839) 171
intelligent design 31, 73–4
'inward glorying' 11, 139–40
Ireland, potato famine (1845–50) 178–9
Isis (Egyptian goddess) 215, 236

Jones, Steve 109

Kant, Immanuel 19, 46n48, 77, 99–100,
    206n12, 207n13
    *On the Theory and Structure of the
    Heavens* (1755) 19n2

Keir, James 230
Kielmeyer, Carl 26–7, 36
Kingsley, Charles 172
Kitchener, Lord 181
Kohn, David 6–7, 9

La Condamine, Charles de 30
Lamarck, Jean-Baptiste de 127, 136, 166
language
    Darwin's use of 8, 13–14, 34, 41, 59
    development of 148–51
Lavoisier, Antoine 53, 230
'leaf archetype' 81–2, 100
Levine, George 163, 193, 194, 199, 211,
    214–15, 219
Linnaeus, Carl 19, 21, 108, 235
Linnean Society 67n2, 168, 250
London Zoo 119
Longinus 139
Lunar Society 53
Lyell, Charles 29–30, 67n2, 95, 165
    Principles of Geology 92–3, 192–3

Mackintosh, James 123, 129, 144, 206
Malaspina, Alessandro 30

Malthus, Daniel 65
Malthus, Thomas 15n19, 27, 46–8, 58, 63,
    64–6, 67, 167, 176
Malvern, Worcestershire 248
Marx, Karl, on Darwin 43n42
materialism, and Erasmus Darwin 221–3
materialism, Romantic 3, 5, 17, 70, 72,
    121, 137–8, 152, 162–3
Matrimonial Causes Act (1857) 171
Mill, John Stuart 143
Miller, Gordon L. 101, 105, 106
Milton, John, *Paradise Lost* 6, 199–200,
    245
Mind, concept of 3
Montagu, Ashley 66
Moore, James (with Adrian
    Desmond), *Darwin* (1992) 1,
    154, 161
'moral sense' theory 123n8
morphology 82

narrative
    in Chambers' *Vestiges* 175–6
    Darwin's use of 6, 8–9, 76, 93, 114,
    116, 191–2, 197, 200, 212n21, 217
Natural History Museum 240
natural selection, and Alfred Wallace 67
natural selection, Darwin's theory of 13,
    21–4, 32–43, 45–6, 60–1
Nature
    oneness of 3, 8, 43–6, 88, 192, 198,
    229
    unity of 3, 16, 20, 21n10, 22–30, 26,
    44–6, 48, 58, 78, 102, 225
    wholeness of 3, 11, 49, 75–6, 98, 102,
    225
Naturphilosophie 1

O'Hear, Anthony 180–1
'One Reality Nature' 75
O'Neill, Michael 131n23, 132

Orange Tip butterfly, example of sexual selection 156
Owen, Richard 4, 35n38
Oxford evolution debate (1860) 180n22

Padel, Ruth, works of:
Darwin: A Life in Poems 239–52
The Rime of the Ancient Naturalist 17
Paley, William 1, 31, 73–4, 123
'Pangenesis hypothesis' 127n16
Patmore, Coventry, *The Angel in the House* 169n9, 212
Peel, Robert 178
phlogiston theory 53
photography 49n51
phrenology 57, 169
Physiocrats 66, 227
Piombo, Sebastiano del, *The Raising of Lazarus* 184–9
plant geography 95–9
Plato 47
Playfair, John 194n3
Pliny the Elder, *Natural History* 29n22
poetry
  and Darwin's moral sense 133, 135, 143–4
  importance to Darwin 191–2, 199–208, 218–19
Pole, Elizabeth 236
Polidori, John 222, 224
polygenists 161n53
Poor Laws 176
Priestley, Joseph 53, 229, 230
Priestman, Martin 227
'Primal Plant' (*Urpflanze*) *see* 'leaf archetype'
primordial forms 4, 36, 79, 110, 115, 155–6, 162, 215, 217
Primordial Soul 142–3
progenitor, common 4n11, 15, 16, 37, 55, 57, 59–60, 59n15, 71, 79, 110, 112–15, 121, 132, 152, 182, 183

progenitors 13, 35, 36, 79, 160, 162, 176
Pythagoras 228–9

Quesnay, François 66, 230

racism, and Victorian values 180–1
'rastro' method, Darwin's use of 7–10, 98–9, 191–9, 218, 239–40, 252
'reflection', Darwin's concept of 119–21
Reform Acts 168, 183
Rhodes, Cecil 181
Richards, Robert 4n10, 5, 70, 99, 131n22
Rio Macaò, South America 26
Romantic imagination 8–10, 67, 73–4, 252–4
Romantic materialism 3, 5, 70, 72, 121, 137–8
Romanticism/ Romantic era 1–4, 66, 70
Rousseau, Jean-Jacques 1, 2, 48, 55, 66
rudimentary forms 13, 45, 113–14, 150–1, 162, 172n15, 174
Ruskin, John 171–2

Schelling, Friedrich 76, 78, 82–3, 87, 99
Schiller, Friedrich von 87
Schlegel, Friedrich 1, 132n26
Scott, Walter 174
Secord, James A. 175
Sedgwick, Adam 4, 25, 135n31
self-consciousness 76–7, 82–3, 120, 131–2
self-help, as a Victorian value 166–7
'self-improvement', Darwin's concept of 61, 63, 64
sexual selection 15–16, 152–62, 169, 211–15, 237
Shakespeare, William 207
Shelley, Mary, *Frankenstein* 147, 222–4
Shelley, Percy Bysshe 222, 223, 224
  'A Defence of Poetry' 8–9, 252–3
slave trade 230–2
slavery, Darwin's opposition to 50, 124, 152

Smiles, Samuel 166
Smith, Adam 147
Social Darwinism 15n19, 170n11
Spinoza, Baruch 27, 59, 75, 101
Stephenson, Robert 166
Stewart, Dugald 11
   Essay on the Sublime 139
Stonehenge 196–7
Stott, Rebecca 121n2
'struggle', Darwin's concept of 5–6, 34, 38–40, 42–3, 58–9, 63, 66, 70–2, 99, 166
   and Karl Marx 43n42
sublime, the 11, 64, 75, 134, 138–42, 186, 200, 240
sympathy
   and Erasmus Darwin 232–4
   and moral reflection 121–5, 138–42, 144, 182n23, 254

Tahiti 217, 236
tattooing 217
taxonomy, modern 19n3
teleological forms 4–5, 39, 61n18, 64, 78, 105, 115–16, 146
'tendency hypothesis' 36–7
Tennyson, Alfred, Lord 181
Thomas, Dylan, *Under Milk Wood* 243
Tierra del Fuega 201
Tollet, Georgina 172n13
Townsend, Joseph 63, 65, 66, 166, 179
transmutation, theory of 12, 24, 32, 66, 67, 71n5, 89n11, 137, 169, 241, 244
   see also evolution
Treak Cliff Cavern, Derbyshire 221
'tree of life' metaphor, Darwin's 33, 40–1, 44–5, 57, 61n18, 67, 72, 77, 79, 99, 108, 151
Trollope, Anthony 181

Uglow, Jenny 235
Uniformitarianism 194n3
utilitarianism 122–3

Van Diemen's Land, Chile 197
Victorian era/ values 168, 211n19, 212
Villa Diodati, Geneva 222
volcanoes, Humboldt's study of 24–5, 53, 90
Volta, Alessandro 51, 53
vorticellae 222–3, 226

Wade's London Review (1845) 175
Wallace, Alfred Russel 67, 156, 167–8, 250
Watt, James 228
'web of affinities', Darwin's concept of 16, 19, 30, 32, 33, 41, 94, 114
'wedges' metaphor, Darwin's 6–7, 59, 71, 241
Wedgwood, Fanny 172n12
Wedgwood, Josiah 166–7, 221, 228, 231
Wedgwood, Josiah II 206n11
Wedgwood family 55n7, 167, 206n11
Weismann, August, 'Germ-Plasm' Theory 128n17
Wellington, Mount 197
Welpley, James Davenport 87
Whewell, William 148
Whitley, Charles 184
Wilberforce, Samuel 180–1
Wilberforce, William 57, 180
Williams, Raymond 227
Withering, William 235
women
   Darwin's attitude to 15, 55n8, 169–71, 212–15
   Victorian attitude to 169–72, 212–15
Wordsworth, William 68, 191
   and 'double consciousness' 208n16

Wordsworth, William, works of:
   The Excursion 9, 130–2, 142
   and grief 205–6
   'Love of Nature Leading to Love of Mankind' (from *The Prelude*) 46–8, 233
   Lyrical Ballads 145, 237, 240n1, 243, 248
   Ode to Duty 206
   Preface to the *Lyrical Ballads* 135–6, 140
   The Prelude 141–2, 144
   'Tintern Abbey' 163, 218–19
workhouses 176

31/1/20

Newport Library and
Information Service